ORIGIN AND EVOLUTION OF EARTH

RESEARCH QUESTIONS FOR A CHANGING PLANET

Committee on Grand Research Questions in the Solid-Earth Sciences

Board on Earth Sciences and Resources

Division on Earth and Life Studies

NATIONAL RESEARCH COUNCIL
OF THE NATIONAL ACADEMIES

THE NATIONAL ACADEMIES PRESS
Washington, D.C.
www.nap.edu

THE NATIONAL ACADEMIES PRESS 500 Fifth Street, N.W. Washington, DC 20001

NOTICE: The project that is the subject of this report was approved by the Governing Board of the National Research Council, whose members are drawn from the councils of the National Academy of Sciences, the National Academy of Engineering, and the Institute of Medicine. The members of the committee responsible for the report were chosen for their special competences and with regard for appropriate balance.

This study was supported by the National Science Foundation, Award No. EAR-0533650; National Aeronautics and Space Administration, Award No. NNH06CE15B, TO #104; U.S. Department of Energy, Award No. DE-FG02-05ER15664; and U.S. Department of Interior / U.S. Geological Survey, Award No. 05HQGR0138. Any opinions, findings, conclusions, or recommendations expressed in this publication are those of the author(s) and do not necessarily reflect the views of the organizations or agencies that provided support for the project.

International Standard Book Number-13:   978-0-309-11717-3 (Book)
International Standard Book Number-10:   0-309-11717-3 (Book)
International Standard Book Number-13:   978-0-309-11718-0 (PDF)
International Standard Book Number-10:   0-309-11718-6 (PDF)
Library of Congress Catalog Number: 2008929776

Additional copies of this report are available from the National Academies Press, 500 Fifth Street, N.W., Lockbox 285, Washington, DC 20055; (800) 624-6242 or (202) 334-3313 (in the Washington metropolitan area); Internet, http://www.nap.edu.

Cover: Selection of scales and disciplines relevant to Earth science. *Top*: Artist's conception of an emerging solar system around the star Beta Pictoris. Courtesy of Lynette R. Cook and the National Aeronautics and Space Administration. *Bottom right*: Outcrop of Neoproterozoic (750-800 million years old) platform carbonates (left) and subjacent interbedded carbonaceous shales and stromatolitic carbonates (right) exposed by receding glacier, northeastern Spitsbergen. Courtesy of Andrew Knoll, Harvard University. *Bottom middle*: A spherical-global view (orthographic projection) of the western hemisphere 105 million years ago. Courtesy of Ronald Blakey, Northern Arizona University. *Bottom left*: Ground motion intensities for a simulated magnitude 7.7 earthquake on the San Andreas Fault in the Los Angeles area. Visualization courtesy of Amit Chourasia and Steve Cutchin, San Diego Supercomputer Center, University of California, San Diego, based on data provided by Kim Olsen and colleagues, Southern California Earthquake Center. *Back*: Repeated images of the crystal structure of stishovite, a mantle mineral that can store water in Earth's interior. Courtesy of Lars Stixrude, University of Michigan.

THE NATIONAL ACADEMIES
Advisers to the Nation on Science, Engineering, and Medicine

The **National Academy of Sciences** is a private, nonprofit, self-perpetuating society of distinguished scholars engaged in scientific and engineering research, dedicated to the furtherance of science and technology and to their use for the general welfare. Upon the authority of the charter granted to it by the Congress in 1863, the Academy has a mandate that requires it to advise the federal government on scientific and technical matters. Dr. Ralph J. Cicerone is president of the National Academy of Sciences.

The **National Academy of Engineering** was established in 1964, under the charter of the National Academy of Sciences, as a parallel organization of outstanding engineers. It is autonomous in its administration and in the selection of its members, sharing with the National Academy of Sciences the responsibility for advising the federal government. The National Academy of Engineering also sponsors engineering programs aimed at meeting national needs, encourages education and research, and recognizes the superior achievements of engineers. Dr. Charles M. Vest is president of the National Academy of Engineering.

The **Institute of Medicine** was established in 1970 by the National Academy of Sciences to secure the services of eminent members of appropriate professions in the examination of policy matters pertaining to the health of the public. The Institute acts under the responsibility given to the National Academy of Sciences by its congressional charter to be an adviser to the federal government and, upon its own initiative, to identify issues of medical care, research, and education. Dr. Harvey V. Fineberg is president of the Institute of Medicine.

The **National Research Council** was organized by the National Academy of Sciences in 1916 to associate the broad community of science and technology with the Academy's purposes of furthering knowledge and advising the federal government. Functioning in accordance with general policies determined by the Academy, the Council has become the principal operating agency of both the National Academy of Sciences and the National Academy of Engineering in providing services to the government, the public, and the scientific and engineering communities. The Council is administered jointly by both Academies and the Institute of Medicine. Dr. Ralph J. Cicerone and Dr. Charles M. Vest are chair and vice chair, respectively, of the National Research Council.

www.national-academies.org

Preface

Over the past four decades, Earth scientists have made great strides in understanding our planet's workings and history. We understand as never before how plate tectonics shapes our planet's surface, how life can be sustained over billions of years, and how geological, biological, atmospheric, and oceanic processes interact to produce climate—and climatic change. Yet at the most basic level, this progress has served principally to lay bare more fundamental questions about Earth. Expanding knowledge is generating new questions, while innovative technologies and new partnerships with other sciences provide new paths toward answers.

The Committee on Grand Research Questions in the Solid-Earth Sciences was established at the request of the U.S. Department of Energy, National Aeronautics and Space Administration, National Science Foundation, and U.S. Geological Survey to frame some of the great intellectual challenges inherent in the study of Earth and other planets. Although many reports have identified research priorities in Earth science, few have cast them as compelling, fundamental science questions. Such "big picture" questions may require decades to answer and research support from many agencies and organizations. The answers to these questions could profoundly affect our understanding of the planet on which we live.

The committee began by drafting "strawman" questions and publishing them for comment in *Eos, Transactions of the American Geophysical Union* (Linn, 2006), on the National Academies website, and in electronic newsletters of the American Geological Institute

and the Association of Women Geoscientists. Written input was also gathered from colleagues. The committee met four times to gather input, discuss community feedback, and write its report.

A small committee cannot hope to have all the expertise needed to cover the broad range of topics discussed in this report. Consequently, the committee solicited essays from colleagues. Of particular note were the essays provided by Greg Beroza, Katharine Cashman, and Kevin Zahnle. Other colleagues who devoted many hours helping the committee sort through ideas include Alan Anderson, Richard Bambach, Katherine Freeman, James Kasting, Barbara Romanowicz, Sean Solomon, and Mary Lou Zoback. The committee is deeply appreciative of their contributions. The committee also thanks the many other individuals who provided input or feedback on the questions: Richard Allen, Paul Barton, Steven Benner, David Bercovici, Robert Berner, Robert Blair, Jr., Gudmundur Bodvarsson, Alan Boss, Gabriel Bowen, Susan Brantley, Douglas Burbank, Frank Burke, Kenneth Caldeira, Richard Carlson, John Chambers, Frederick Colwell, Kevin Crowley, Gedeon Dagan, Andrew Davis, William Dickinson, William Dietrich, David Diodato, Bruce Doe, Robert Dott, Jr., Benjamin Edwards, Peter Eichhubl, Michael Ellis, W. Gary Ernst, Douglas Erwin, Rodney Ewing, Fredrick Frey, Arthur Goldstein, Linda Gundersen, David Halpern, Wayne Hamilton, T. Mark Harrison, John Hayes, James Head, Michael Hochella, Jr., Vance Holliday, Richard Iverson, A. Hope Jahren, Raymond Jeanloz, Gerald Joyce, Joseph Kirschvink, John LaBrecque, Thorne Lay, Antonio Lazcano,

Cin-Ty Lee, William Leeman, Jonathan Lunine, Ernest Majer, Michael Manga, Anthony Mannucci, William McDonough, Dan McKenzie, Marcia McNutt, H. Jay Melosh, Peter Molnar, Isabel Montanez, Alexandra Navrotsky, Shlomo Neuman, Gary Olhoeft, Carolyn Olson, Peter Olson, Minoru Ozima, Nazario Pavoni, Donald Porcelli, Jonathan Price, Steven Pride, George Redden, Paul Renne, Robin Reichlin, Mark Richards, Daniel Schrag, Norman Sleep, D. Kip Solomon, Gerilyn Soreghan, Frank Spear, Gary Sposito, Steven Stanley, Ross Stein, Robert Stern, David Stevenson, Lynn Sykes, Jack Szostak, Thomas Tharp, Leon Thomsen, Oliver Tschauner, Terry Tullis, Greg Valentine, Richard Von Herzen, Joseph Wang, James Whitcomb, Raymond Willemann, M. Gordon Wolman, Nicholas Woodward, Eva Zanzerkia, Xiaobing Zhou, and Herman Zimmerman.

Donald J. DePaolo, *Chair*

Acknowledgment of Reviewers

This report has been reviewed in draft form by individuals chosen for their diverse perspectives and technical expertise, in accordance with procedures approved by the National Research Council's Report Review Committee. The purpose of this independent review is to provide candid and critical comments that will assist the institution in making its published report as sound as possible and to ensure that the report meets institutional standards for objectivity, evidence, and responsiveness to the study charge. The review comments and draft manuscript remain confidential to protect the integrity of the deliberative process. We wish to thank the following individuals for their participation in the review of this report:

Douglas Erwin, Smithsonian Institution
Jonathan Fink, Arizona State University
Jeffrey Freymueller, University of Alaska
Russel Hemley, Carnegie Institution of
   Washington
Thomas Jordan, University of Southern
   California
Louise Kellogg, University of California, Davis
Rosamond Kinzler, American Museum of
   Natural History

Jay Melosh, University of Arizona
Franklin Orr, Stanford University
Norman Sleep, Stanford University
Steven Stanley, University of Hawaii
David Stevenson, California Institute of
   Technology
Robert van der Hilst, Massachusetts Institute of
   Technology

Although the reviewers listed above have provided many constructive comments and suggestions, they were not asked to endorse the conclusions or recommendations nor did they see the final draft of the report before its release. The review of this report was overseen by Marcia McNutt, Monterey Bay Aquarium Research Institute. Appointed by the National Research Council, she was responsible for making certain that an independent examination of this report was carried out in accordance with institutional procedures and that all review comments were carefully considered. Responsibility for the final content of this report rests entirely with the authoring committee and the institution.

Contents

Summary

Modern science has its roots in fundamental questions about the origins of Earth and life. These grand questions are recorded in texts of the ancient Greeks, who laid the foundations of Earth science and whose language provides many of its terms. Analytical approaches to answering these questions date back to the 16th century for planetary science and the 18th century for geological science. Perhaps the first, and certainly one of the most controversial, of the more modern grand research questions in geology came from observations of sedimentary rocks. The thickness of sedimentary beds, their variable character and structures, and the presence of fossils within them led James Hutton to conclude that Earth must be very old (Hutton, 1788). The age of Earth became the ultimate grand question of the time. But not until almost 200 years later—after it was established that matter was made of atoms, that atoms had nuclei, and that some of those nuclei were unstable to radioactive decay—was it possible to establish the scale of geological time. The first accurate measurement of Earth's age, 4.55 billion years, made in the mid-1950s (Patterson, 1956), was a major step in establishing a timescale for Earth, for life, and for the Universe.

Until the 1960s, geological science was built almost entirely on the study of rocks and landforms on the continents; little was known about the seafloor. The grand research questions of the early 20th century were heavily influenced by this continent-centric view, as well as by a focus on mineral and water resources and discoveries in paleontology. There were grand questions about how volcanoes, mountain ranges, and sedimentary basins were created; why mineral deposits and petroleum deposits formed where and when they did; how fast mountains were built and eroded away; why fossils first became abundant only 500 million years ago; and what caused ice ages and earthquakes. An additional tantalizing question was why the Atlantic coastlines of South America and Africa looked like they were pieces of a puzzle that might once have been joined together.

This seemingly unconnected set of grand questions of the mid-20th century were largely organized and linked by the advent of plate tectonics theory. In just half a decade, between 1963 and 1968, spurred largely by the first observations of the magnetism and depth of the seafloor, a grand picture of the dynamic behavior of the planet emerged. It was deduced that Earth's surface consists of a dozen or so irregular, stiff plates that move a few centimeters per year and that the boundaries of these plates are the locations of earthquakes, volcanoes, and mountain ranges. The plate movements are connected to a planetwide system of solid-state convection deep within Earth, an idea that was inconceivable to most geologists a decade before.

The plate tectonics model, including its corollaries of mantle convection, seafloor spreading, and continental drift, not only explained the pattern of earthquakes, volcanoes, and mountain ranges but also eventually provided possible mechanisms to create the continents and seafloor, to gradually shift Earth's climate over geological time, and to influence the course of biological evolution. Toward the end of this watershed period of the 1960s, the United States landed the first astronauts on the Moon, who brought back rock samples that provided a glimpse of another planetary body much different from Earth. This new perspective ushered in the modern era where Earth is viewed as a planet and its constitution, history, and character are compared to those of other planets.

In 1980 another breakthrough came from evidence that Earth was struck by a large meteoroid 65 million years ago and that the impact probably caused the extinction of dinosaurs and many of the other living things on the planet at the time (Alvarez et al., 1980). Within a few years it became evident that some meteorites found on Earth came from Mars (Bogard and Johnson, 1983). These two developments underscored the idea, which had begun with studies of impact craters on Earth and the Moon, that Earth must be viewed in its astronomical context; for example, life could be terminated by uninvited extraterrestrial objects or imported from other Solar System planets!

Over the past 20 years the transformation of Earth science has continued. Major advances in technology that allow Earth to be observed much better at both large and small scales, continuing planetary exploration, and advanced computing have all contributed. We can now see into minerals and discern individual atoms, measure the properties of rocks at the immense pressures and temperatures inside Earth, watch continents drift and mountains grow in real time, and understand how organisms evolve and interact with Earth based on their DNA. We have also been able to extract new information from meteorites that tells us about how planets form and even about how the interiors of stars work. Armed with new tools, Earth science is turning to the deeper fundamental questions—the origin of Earth; the origin of life; the structure and dynamics of planets; the connections between life, climate, and Earth's interior; and what the Earth may hold for humankind in the future.

SCOPE AND PURPOSE OF THIS REPORT

At the request of the U.S. Department of Energy, the National Science Foundation, the U.S. Geological Survey, and the National Aeronautics and Space Administration, the National Academies established a committee to propose and explore grand Earth science questions being pursued today. The charge to the committee, given below, provided unusual freedom in the selection of topics, without regard to agency-specific issues, such as mission relevance and implementation.

> The committee will formulate a short list of grand research questions driving progress in the solid-Earth sciences. The research questions will cover a variety of spatial scales and temporal scales, from subatomic to planetary and from the past (billions of years) to the present and beyond. The questions will be written in a clear, compelling way and will be supported by text and figures that summarize progress to date and outline future challenges. This report will not discuss implementation issues (e.g., facilities, recommendations aimed at specific agencies) or disciplinary interests.

Our response to this charge has been to attempt to capture the scope and aspirations of what might best be referred to as geological and planetary science, which is another way of saying solid-Earth science. Research in this area draws on nearly every scientific discipline. However, research questions that are mainly the domain of other subdisciplines of Earth science—such as ocean, atmospheric, or space science—are discussed to the extent they are linked to solid-Earth science.

The committee began by developing criteria for what constitutes a "grand" question. Our definition of grand questions was partly determined by the small number requested in the charge, which led us to aim for 7 to 10 questions, and partly by a desire for the questions to meet at least two of the following criteria:

- it transcends the boundaries of a narrow subfield of geological and planetary science;
- it deals with eternal issues, such as the origins of Earth and life;
- it is connected with phenomena that have significant impact on human well-being.

Our ultimate objective was to capture in this series of questions the essential scientific issues that constitute the frontier of Earth science at the start of the 21st century. It is our hope that these questions and our descriptions of them are as compelling as we believe the science to be and that this short report is useful to those who would like to understand more about where Earth science stands, how it got there, and where it might be headed. We have attempted to make the text accessible to managers of scientific programs, graduate students, and colleagues in sister disciplines who have the technical or scientific background needed to comprehend what is discussed.

Our most difficult problem in selecting the grand questions was to distill from a large number of topics and questions the "most worthy" candidates. To do so the committee canvassed the broad geological community and deliberated in meetings and telephone conferences. After arriving at 10 grand questions, the committee set about writing, as well as soliciting written contributions from other scientists. Some of our questions present truly awesome challenges and may not be fully understood for decades, if ever. Others seem more tractable, and significant progress may be made in a matter of years. Overall, we have included most of what the committee regards as the important issues and also most of what was suggested by the respondents to our canvassing effort. There was, in fact, a fair degree of consensus about what constitutes a grand question and which ones should be included here.

GRAND RESEARCH QUESTIONS FOR THE 21ST CENTURY

Although we started by simply identifying the overarching questions we believe to be driving modern Earth science, we found that these questions can be grouped into four broad themes. These themes constitute the four chapters of the report, and within each chapter are descriptions of the grand questions. Chapter 1 deals with origins—the origin of Earth and other Solar System planets, Earth's earliest history, and the origin of life. Chapter 2 treats the workings of Earth's interior and its surface manifestations and includes a question on material properties and their fundamental role in Earth processes. Chapter 3 addresses the habitability of the surface environment—climate and climate change and Earth–life interactions. Chapter 4 focuses on geologica10

hazards and Earth resources—earthquakes and volcanoes and modern environmental issues associated with water and other fluids in and on Earth.

The following is a summary of the 10 grand research questions identified by the committee:

1. How did Earth and other planets form? The Solar System, with its tantalizing geometric patterns and its wide variety of planets and moons, presents intriguing questions that become more nuanced as we make new observations from spacecraft and more exacting measurements on meteorites. While it is generally agreed that the Sun and planets all coalesced out of the same nebular cloud, it is still not known how Earth obtained its particular chemical composition, at least not in enough detail to understand its subsequent evolution or why the other planets ended up so different from ours and from each other. Earth, for example, has retained a life-giving inventory of volatile substances, including water, but Earth is far different from every other planet in this regard. Advanced computing capabilities are enabling development of more credible models of the early Solar System, but further measurements of other Solar System bodies and extrasolar planets and objects appear to be the primary pathway to furthering our understanding of the origin of Earth and the Solar System.

2. What happened during Earth's "dark age" (the first 500 million years)? It is now believed that in the later stages of Earth's formation, a Mars-sized planet collided with it, displacing a huge cloud of debris that became our Moon. This collision added so much heat to Earth that the entire planet melted. Little is known about how this magma soup differentiated into the core, mantle, and lithosphere of today or how Earth developed its atmosphere and oceans. The so-called Hadean Eon is a critical link in our understanding of planetary evolution, but we have little information about it because there are almost no rocks of this age preserved on Earth. Clues about this time period are accumulating, however, as we learn more about meteorites and other planets and extract new information from ancient crystals of zircon on Earth.

3. How did life begin? The origin of life is one of the most intriguing, difficult, and enduring questions in science. Because life in the Solar System arose billions of years ago, some of the most fundamental questions about its origin are geological. Our knowledge of the materials from which life originated, and where, when, and in what form it first appeared, stems from geological investigations of rocks and minerals that represent the only remaining evidence. When life first arose, the conditions at Earth's surface may have been much different than today's, and one critical challenge is to develop an accurate picture of the physical environments and the chemical building blocks available to early life. The quest to establish the origin of life is inherently multidisciplinary, spanning organic chemistry, molecular biology, astronomy, and planetary science, as well as geology and geochemistry. There is growing interest in studying Mars, where there is a sedimentary record of early planetary history that predates the oldest Earth rocks and other star systems where planets have been detected.

4. How does Earth's interior work, and how does it affect the surface? As planets age, they gradually cool, and this causes them to move through stages where their internal processes, their atmospheres, and their surface processes are gradually changing. The primary means by which heat is moved from the interior to the surface is planetwide solid-state and liquid convection. Although we know that the mantle and core are in constant convective motion, we can neither precisely describe these motions today nor calculate with confidence how they were different in the past. Core convection produces Earth's magnetic field, which may have had an important influence on surface conditions. Mantle convection is the cause of volcanism, seafloor generation, and mountain building, and materials like water and carbon are constantly exchanged between Earth's surface and its deep interior. Consequently, without detailed knowledge of Earth's internal processes we cannot deduce what Earth's surface environment was like in the past or predict what it will be in the future.

5. Why does Earth have plate tectonics and continents? The questions regarding plate tectonics now have less to do with the soundness of the theory than with why Earth has plate tectonics in the first place and how closely it is related to other unique aspects of Earth—the abundant water, the existence of continents and oceans, and the existence of life. We do not know whether it is possible to have one aspect without the others or how they are interdependent. The existence and persistence of continental crust present problems as fundamental as those of plate tectonics. Continental crust makes the planet habitable by nonmarine life, and weathering of its surface plays a role in regulating

Earth's climate. But we still do not know when continents first formed, how they are preserved for billions of years, or exactly how they evolved to be what they are like today. New data and observations indicate that climate and erosion play a fundamental role in building and shaping mountain ranges and thus are fundamental to the formation as well as the destruction of continental crust.

6. How are Earth processes controlled by material properties? Deciphering the secrets of the rock record on Earth and other planets begins with the understanding of large-scale geological processes. The keys to understanding these processes are the basic physics and chemistry of planetary materials. The high pressures and temperatures of Earth's interior, the enormous size of Earth and its structures, the long expanse of geological time, and the vast diversity of materials and properties all present special challenges. These challenges are being met with new research tools based on synchrotron radiation, new measurements and simulation capabilities for large domains and heterogeneous materials, and quantum mechanics-based calculations of material properties under extreme conditions. New research areas are developing around the study of natural nanoparticles and the mediation of chemical processes by microorganisms.

7. What causes climate to change—and how much can it change? Global climate conditions have been favorable and stable for the past 10,000 years, but we also know from geological evidence that momentous changes in climate can occur in periods as short as decades or centuries. Yet despite the numerous factors that can change climate, from the slowly changing luminosity of the Sun to the building of new mountain ranges and changes in atmospheric composition, Earth's surface temperature seems to have remained within relatively narrow limits for most of the past 4 billion years. How does it remain well regulated in the long run, even though it can change so abruptly? Recent discoveries have highlighted periods of Earth history when the climate was extremely cold, was extremely hot, or changed especially quickly. Understanding these special conditions may lead to new insights about Earth's climate, as will new geochemical observations made on ancient sedimentary rocks and improved models for the climate system that will eventually enable us to predict the magnitude and consequences of climate changes.

8. How has life shaped Earth—and how has Earth shaped life? Earth scientists have a tendency to view Earth's geological evolution as a fundamentally inorganic process. Life scientists, in the same spirit, tend to regard the evolution of life as a fundamentally biological issue. Yet the development of life has clearly been influenced by the conditions of Earth's surface, while Earth's surface has been influenced by the activities of life forms. The atmosphere would not contain oxygen if it were not for life, and the presence of oxygen has enabled other types of life to evolve. We know that geological events and meteoroid impacts have caused massive extinctions in the past and influenced the course of evolution. But the exact ties between geology and evolution are still elusive. On the modern Earth we are interested in the role of life in geological processes like weathering and erosion. And we seek to understand how life may have manifested itself and left traces preserved in the geological records of other planets.

9. Can earthquakes, volcanic eruptions, and their consequences be predicted? Thanks largely to sensitive new instrumentation and better understanding of causes, geologists are moving toward predictive capabilities for volcanic eruptions. For earthquakes, progress has been made in long-term forecasts, but we may never be able to predict the exact time and place an earthquake will strike. Continuing challenges are to deepen our understanding of how fault ruptures start and stop, to improve our simulations of how much shaking can be expected near large earthquakes, and to increase the warning time once a dangerous earthquake

begins. Studies of volcanic activity have entered a new era as a result of real-time seismic, geodetic, and electromagnetic probes of active subsurface processes. But it remains a challenge to integrate such real-time data with field studies of volcanoes and laboratory studies of volcanic materials. The ultimate objective is to develop a clear picture of the movement of magma, from its sources in the upper mantle to Earth's crust, where it is temporarily stored, and ultimately to the surface where it erupts.

10. How do fluid flow and transport affect the human environment? Good management of natural resources and the environment requires knowledge of the behavior of fluids, both below ground and at the surface. The major scientific objectives are to understand how fluids flow, how they transport materials and heat, and how they interact with and modify their surroundings. New experimental tools and field measurement techniques, plus airborne and spaceborne measurements, are offering an unprecedented view of processes that affect both the surface and the subsurface. But we still have difficulty determining how subsurface fluids are distributed in heterogeneous rock and soil formations, how fast they flow, how effectively they transport dissolved and suspended materials, and how they are affected by chemical and thermal exchange with the host formations. Much better models of streamflow and associated erosion and transport are needed if we are to accurately assess how human impacts and climate change affect landscape evolution and how these effects can be managed to sustain ecosystems and important watershed characteristics. The ultimate objective—to produce mathematical models that can predict the performance of natural systems far into the future—is still out of reach but critical to making informed decisions about the future of the land and resources that support us.

1

Origins

The modern study of Earth is ultimately rooted in humankind's desire to understand its origins. Although it was once assumed that intelligent life was unique to Earth, we have now gained an appreciation that even though it may not be unique, the existence of advanced life on planets may well be uncommon. None of the other planets of the Solar System are presently suitable for the complex life forms that exist on Earth, and we have yet to identify other stars that have planets much like Earth. Although the odds are good that there is other life in our galaxy, this inference has not been confirmed.

Considering the apparent rarity of Earth-like life, it is natural to want to understand what went into making Earth suitable for life and how life arose. Pursuing these questions leads us to fundamental issues about how stars and planets form and evolve and to questions about how the modern Earth works, from the innermost core to the atmosphere, oceans, and land surface. This chapter presents three questions related specifically to origins—one regarding the origin of Earth and other planets and one regarding the origin of life. These two questions are separated by a third that deals with Earth's earliest history: the 500 million to 700 million years between the time of the origin of the Solar System and the oldest significant rock record preserved on Earth. During this early, still poorly understood, stage of Earth's development, tremendous changes must have taken place, accompanied by myriad catastrophic events, all leading ultimately to a setting in which life could develop and eventually thrive.

QUESTION 1: HOW DID EARTH AND OTHER PLANETS FORM?

One of the most challenging and relevant questions about Earth's formation is why our planet is the only one in the Solar System with abundant liquid water at its surface and abundant carbon in forms that can be used to make organic matter. This question is part of a broader set: why the inner planets are rocky and the outer planets are gaseous; how the growth and orbital evolution of the outer planets influenced the inner Solar System; why all of the largest planets are so different from one another; and how typical our Solar System is within the Milky Way galaxy. Although these questions are longstanding, the answers are only now emerging from new insights provided by astronomy, isotopic chemistry, Solar System exploration, and advanced computing. And although we know in general how to make a planet like Earth—starting with some stardust and allowing gravity, radiation, and thermodynamics to do their parts—our answers often serve only to refine our questions. For example, the details of Earth's chemical composition—such as how much of the heat-producing elements uranium, thorium, and potassium it contains; how much oxygen and carbon it contains; and how it came to have its particular allotment of noble gases and other minor constituents—turn out to be critical to models of Earth's geological processes and, ultimately, to understanding why Earth has remained suitable for life over most of its history.

How Do Planets Form Around Stars?

We do not know how unique or unusual the Solar System is, but observations of other planetary systems are providing new ideas for how planets form and evolve. Astronomical observations of star-forming regions and young stars, together with hydrodynamic models of star formation, support the conclusion that stars—including the Sun—form by the gravitational collapse of a molecular cloud core composed of materials manufactured and reprocessed in many earlier generations of stars. Because the typical molecular cloud is rotating at the time of collapse, the developing star is surrounded by a rotating disk of gas and dust. Most disks around young stars, as viewed through telescopes, are approximately 99 percent gas and 1 percent dust, but even that

small proportion of dust makes the disks opaque at visible wavelengths (Figures 1.1 and 1.2). Gas-giant planets, such as Jupiter and Saturn in our system, are believed to form in such circumstellar disks, but direct astronomical observations of planets forming have not yet been made.

Observations of planets around other nearby stars with masses similar to the Sun indicate that planet formation is a common outcome of star formation, but no star has yet been observed with a system of planets that looks anything like the Solar System. Over 200 extrasolar planets have been discovered by several indirect techniques (e.g., radial velocity of the host star, stellar transit, and microlensing) (Butler et al., 2006; <www.exoplanets.org>). Multiple planets are known to orbit some two dozen stars. The vast majority of these

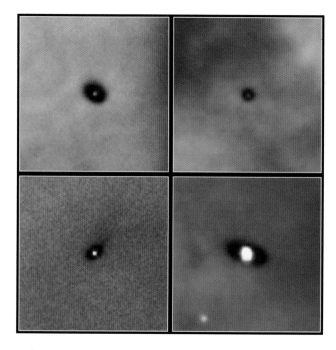

FIGURE 1.1 Hubble Space Telescope images of four protoplanetary disks around young stars in the Orion nebula, located 1,500 light-years from the Sun. The red glow in the center of each disk is a newly formed star approximately 1 million years old. The stars range in mass from 0.3 to 1.5 solar masses. Each image is of a region about 2.6 × 10^{11} km (400 AU) across and is a composite of three images taken in 1995 with Hubble's Wide Field and Planetary Camera 2 (WFPC2), through narrow-band filters that admit the light of emission lines of ionized oxygen (represented by blue), hydrogen (green), and nitrogen (red). SOURCE: Mark McCaughrean, Max Planck Institute for Astronomy; C. Robert O'Dell, Rice University; and the National Aeronautics and Space Administration, <http://hubblesite. org/gallery/album/nebula_collection/pr1995045b/>.

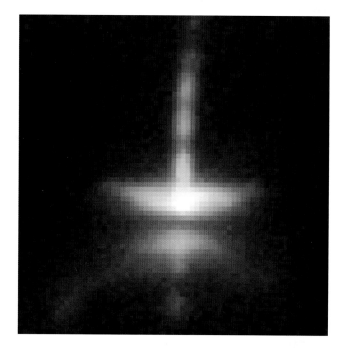

FIGURE 1.2 Hubble Space Telescope WFPC2 image of Herbig-Haro 30, a prototype of a young (approximately 1-million-year-old) star surrounded by a thin, dark disk and emitting powerful bipolar jets of gas. The disk extends about 6 × 10^{10} km from left to right in the image, dividing the edge-on nebula in two. The central star is hidden from direct view, but its light reflects off the upper and lower surfaces of the flared disk to produce the pair of reddish nebulae. The gas jets, shown in green, are driven by accretion. SOURCE: Chris Burrows, Space Telescope Science Institute; John Krist, Space Telescope Science Institute; Kare Stapelfeldt, Jet Propulsion Laboratory; and colleagues; the WFPC2 Science Team; and the National Aeronautics and Space Administration, <http://hubblesite. org/gallery/album/entire_collection/pr1999005c/>.

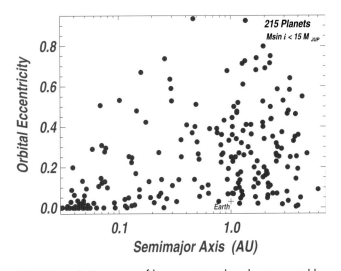

FIGURE 1.3 Summary of known extrasolar planets sorted by distance from host star and orbital eccentricity. All of the planets in the Solar System have eccentricities of 0.2 or less. SOURCE: Courtesy of Geoffrey Marcy, University of California, Berkeley. Used with permission.

planets are thought to be gas giants on the basis of their masses and densities. Presumably, more gas giants are observed because they are large, and large planets are much easier to detect, leaving open the question of how many terrestrial planets remain hidden from Earth in distant planetary systems. A few "super-Earths," with masses of several to 10 Earth masses, may be terrestrial planets, but no measurements of the radius or density of these objects has confirmed this. Gas-giant planets appear to be more likely with stars that have proportions of heavier elements (heavier than H, He, and Li) as high as the Sun (Fischer and Valenti, 2005), suggesting that heavy-element concentrations in the circumstellar disk influence the rate or efficiency of planet formation.

Measurements of the masses, orbital distances, and orbital eccentricities (Figure 1.3) of extrasolar planets provide clues about processes that may help determine what the final planetary system looks like. A particularly interesting class of planets, that of gas-giant planets in orbits extremely close to (less than 0.1 AU)[1] their host stars—sometimes called "hot Jupiters"—are significant because models have been unable to account

for why they form so close to the star (Butler et al., 2006). These hot Jupiters are thought to be telling us that large planets can drift inward toward their star as they form. Models also suggest that planets can under some circumstances drift away from the star, so the ultimate location of the planets may have little to do with where they originally formed. Extrasolar planets more than a few tenths of an AU distant from their host star often have quite eccentric orbits, which contrasts with the Solar System where all of the planets except Mercury have nearly circular orbits.

How Did the Solar System Planets Form?

The Solar System is composed of radically different types of planets. The outer planets (Jupiter, Saturn, Uranus, and Neptune) are distinguished from the inner planets by their large size and low density. The outer planets are the primary products of the planet formation process and comprise almost all of the mass held in the planetary system. They are also the types of planet that are most easily recognized orbiting other stars. The inner planets (Mercury, Venus, Earth, and Mars) are composed mostly of rock and metal, with only minor amounts of gaseous material. There are "standard models" for the formation of both types of planets, but they have serious deficiencies and large uncertainties.

According to the standard model for outer-planet formation, the formation of giant planets starts with condensation and coalescence of rocky and icy material to form objects several times as massive as Earth. These solid bodies then attract and accumulate gas from the circumstellar disk (Pollack et al., 1996). The two largest outer planets, Jupiter and Saturn, seem to fit this model reasonably well, as they consist primarily of hydrogen and helium in roughly solar proportions, but they also include several Earth masses of heavier elements in greater than solar proportions, probably residing in a dense central core. Uranus and Neptune, however, have much lower abundances of hydrogen and helium than Jupiter and Saturn and have densities and atmospheric compositions consistent with a significant component of outer Solar System ices.

An alternative to the standard model is that the rock and ice balls are not needed to induce the formation of gas-giant planets; they can form directly from the gas and dust in the disk, which can collapse under

---

[1]The astronomical unit, or AU, is a unit of length nearly equal to the semimajor axis of Earth's orbit around the Sun, or about 150 million km.

its own gravity like miniversions of the Sun (Boss, 2002). In this model the excess abundances of heavy elements in Jupiter and Saturn would have been acquired later by capture of smaller rocky and icy bodies. This model, however, does not account well for the compositions of Uranus and Neptune, which do not have very much gas. Other important questions about the outer planets are when they formed and the extent to which they may have drifted inward or outward from the Sun during and after formation. Where the outer planets were and when is important for understanding how the inner planets formed.

The primary difference between the inner and outer planets (rock versus gas and ice) is thought to reflect the temperature gradient in the solar nebula. Temperatures were relatively high (>1000 K) near the developing Sun, dropping steadily with distance. Near the Sun, mainly silicates and metal would have condensed from the gas (so-called refractory materials), whereas beyond the asteroid belt, temperatures were low enough for ices (i.e., water, methane, ammonia) containing more volatile elements to have condensed, as well as solid silicates. It was once thought that as the nebula cooled, solids formed in a simple unidirectional process of condensation. We now know that solids typically were remelted, reevaporated, and recondensed repeatedly as materials were circulated through different temperature regimes and variously affected by nebular shock waves and collisions between solid objects. Important details of the temperatures of the solar nebula, however, are still uncertain, including such significant issues as peak temperatures, how long they were maintained, and how temperature varied with distance from the Sun and from the midplane of the disk. Defining these conditions is an important part of understanding how the chemical compositions of the planets and meteorites came to be.

The standard model for the formation of the inner planets is somewhat more complicated than the model for outer-planet formation and is based largely on theory and anchored in information from meteorites and observations of disks around other stars (Chambers, 2003). The model strives to explain how a dispersed molecular cloud with a small amount of dust could evolve into solid planets with virtually no intervening gas and how the original mix of chemical elements in the molecular cloud was modified during that evolution. Significant unknowns are how long the process took, how solid materials were able to coagulate into progressively larger bodies, and how and when the residual gas was dissipated. The time for centimeter-sized solid objects to form at Earth's distance from the Sun, according to the standard model, might have been as short as 10,000 years. These small solid objects were highly mobile, pulled Sun-ward large distances by the Sun's gravity as a result of drag from the still-present H-He gas. Submeter-sized objects were also strongly affected by turbulence in the gas.

A particular deficiency of the standard model is its inability to describe the formation of kilometer-sized bodies from smaller fragments. The current best guess is that the dust grains aggregated slowly at first, and growth accelerated along with object size as small objects were embedded into larger ones (Weidenschilling, 1997). The aggregation behavior of objects greater than a kilometer in size is better understood: they are less affected by the presence of gas than are smaller pieces, and their subsequent evolution is governed by mutual gravitational attractions. Growth of still larger bodies, or planetesimals, from these kilometer-sized pieces should have been more rapid, especially at first. Gravitational interactions gave the largest planetesimals nearly circular and coplanar orbits—the most favorable conditions for sweeping up smaller objects. This led to runaway growth and formation of Moon- to Mars-sized planetary embryos. Growth would have slowed when the supply of small planetesimals was depleted and the embryos evolved onto inclined, elliptical orbits. Dynamical simulations based on statistical methods and specialized computer codes are finding that a number of closely spaced planetary embryos are likely to have formed about 100,000 years after planetesimals appeared in large numbers (e.g., Chambers, 2003).

The later stages of planet formation took much longer, involved progressively fewer objects, and hence are less predictable (Figure 1.4). The main phase of terrestrial planet formation probably took a few tens of millions of years (Chambers, 2004). The final stages were marked by the occasional collision and merger of planetary embryos, which continued until the orbits of the resulting planets separated sufficiently to be protected from additional major collisions.

Although there are four terrestrial planets, models suggest that the number could easily have been three

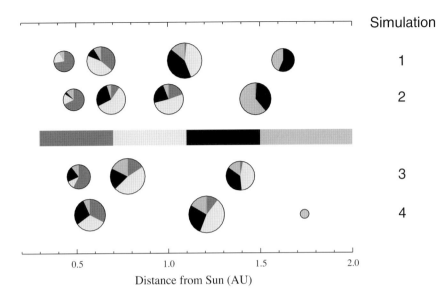

Simulation

1

2

3

4

FIGURE 1.4 Results of four representative numerical simulations of the final stage of accretion of the terrestrial planets. The segments in each pie show the fraction of material originating from the four regions of the solar nebula shown by the shades of gray, and the size of the pie is proportional to the volume of each planet. In each simulation the largest planet has a size similar to Earth's, but there can be either two or three other planets, and the sizes vary. The planets typically receive material from all four zones, with preference for the zones closest to their final orbit location. SOURCE: Chambers (2004). Copyright 2004 by Elsevier Science and Technology Journals. Reprinted with permission.

Distance from Sun (AU)

or five, and they would have been at different distances from the Sun (Figure 1.4). Tidal interactions with nebular gas may have caused early-formed inner planets to migrate inward substantially while they were forming, and several planets may have been lost into the Sun before the gas dispersed (McNeil et al., 2005). The fact that there are no rocky planets beyond Mars is likely a consequence of the presence of the giant planets, particularly Jupiter. The large mass and strong gravitational pull of Jupiter probably prevented the formation of additional rocky planets in the region now occupied by the asteroid belt by disrupting the orbits of bodies in that region before they could form a large planet. Jupiter and Saturn also sent objects from the asteroid belt either out of the Solar System or spiraling into the inner-planet region where they became parts of the planets forming there or fell into the Sun. The asteroids represent the 0.01 percent of material that survived this process.

What Do Meteorites Say About the Origin of Earth?

Earth has undergone so much geological change that we find little evidence in rocks about its origin or even its early development (Question 2). Many meteorites, on the other hand, were not affected by the high-temperature processing that occurs in planetary interiors. They are fragments of, or soil samples from, miniplanets that formed in what is now the asteroid

belt just as the Solar System was starting out. Thus, they preserve significant clues about the state of the Solar System when the planets were forming (Figure 1.5). For this reason, studies of meteorites play a major role in helping us understand Earth's origin.

One gift of meteorites is to reveal the age of the Solar System. Precise radiometric dating of high-temperature inclusions within meteorites shows that the first solid objects in our home system formed 4,567 million years ago (see Box 1.1). We also know that shortly thereafter planetesimals of rock and metal formed and developed iron-rich cores and rocky crusts (see Question 2). Some meteorites are chemically like the Sun (for elements other than H, He, Li, C, N, O, and noble gases), and some of these same meteorites contain tiny mineral grains of dust that survived from earlier generations of stars (see Box 1.2). Other meteorites are parts of small planetary bodies that experienced early volcanism and that were later broken up by collisions. Beyond these clues, meteorites fall short of providing all the information needed to understand Earth, partly because most of them formed far from the Sun (the main asteroid belt is between Mars and Jupiter), and the relationship between meteorites and planets is not fully understood. The systematic collection of well-preserved samples from Antarctica has greatly expanded the number of meteorites available for study and has yielded rarities such as meteorites from Mars and the Moon.

Beyond what they tell us about Earth, meteorites

Allende

FIGURE 1.5 The Allende meteorite, a carbonaceous chondrite, is a mixture of CAIs (calcium-, aluminum-rich inclusions; larger irregularly shaped light-colored objects) and chondrules (round light-colored objects) in a dark-colored matrix of minerals and compounds. The CAIs and chondrules are a high-temperature component that formed and were in some cases reprocessed at temperatures above 1000°C. SOURCE: Hawaii Institute of Geophysics and Planetology. Used with permission.

provide a benchmark for understanding the composition of the Sun and even the Universe as a whole. Most of the visible mass of the Universe, and almost all stars, is composed primarily of hydrogen and helium made during the Big Bang. The rest of the elements—the "heavier" ones with more protons and neutrons in their nuclei—were produced by nucleosynthesis, or thermonuclear reactions within stars. Most nucleosynthesis happens in big stars. These massive stars last only about 10 million to 20 million years before they explode as supernovae. The new elements they make, before and during the explosion, are thrown back into space where they are later recycled into new stars. In the approximately 10 billion years between the origin of the Universe and the origin of the Solar System, hundreds of generations of massive stars have exploded, and over this long period about 1 percent (by weight) of the original H and He has been converted to heavier elements. Meteorites give us the most detailed information about the abundances of these heavier elements.

Meteorites tell us still more about the formation of the Solar System out of the nebular disk. The abundance of heavy elements in the Sun is known

moderately well from spectroscopic data. The planets, however, formed from the nebular disk, so it is important to know whether the disk had the same composition as the Sun, and whether it was homogeneous or varied significantly in composition, perhaps with radial distance from the proto-Sun. The standard model for the composition of the solar nebula is based on studies of a class of meteorites called chondrites (Figure 1.5). Chondrites, the commonest type of meteorites, are stony bodies formed from the accretion of dust and small grains that were present in the early Solar System. They are often used as reference points for chemicals present in the original solar nebula. The most primitive of these objects—those least altered by heat and pressure—are carbonaceous chondrites, whose chemical compositions match that of the Sun for most elements. The relative amounts of elements and their isotopes can be measured much more precisely on meteoritic materials than by solar spectroscopy, so chondritic meteorites play a special role in helping to understand both Earth and nucleosynthesis in our galaxy. Because chondritic elemental abundances look similar to those of the Sun, the disk likely had about the same composition as the Sun.

What Is the Chemical Composition of Earth?

The most critical question related to the formation of Earth is why the planet has its particular chemical makeup. Although we know quite a lot about this issue, a key unanswered question is the origin of Earth's water. Earth, like other objects forming near the Sun, is thought to have formed mainly as a relatively high-temperature partial condensate from a gas of solar composition. The uncondensed gas containing water, carbon, and other volatile elements was swept away by the early solar wind or by ultraviolet radiation pressure. Much of the volatile elements that might have been incorporated into the early Earth is thought to have been lost during the intense heating of the Hadean Eon (Question 2).

It has been suggested that the giant planets can pluck materials from the asteroid belt region and throw them in toward the Sun. Objects beyond Mars would have formed in a cooler part of the solar nebula and hence would likely have contained more volatile compounds. Studies of asteroids indicate that meteorites

BOX 1.1 Time, the Early Solar System, and the Age of Earth

The initial events in the formation of the Sun, meteorites, and Earth and other planets took place in only a few million years, about 4,567 million years ago (Ma). Documenting early Solar System timescales is therefore a substantial challenge. Advances in geochronological techniques are beginning to enable the sequence of events to be discerned.

The most primitive chondritic meteorites contain inclusions made up of minerals that condense at high temperature from a gas of solar nebula composition. These objects, called calcium-aluminum-rich inclusions (CAI), have recently been precisely dated using the decay of uranium to lead, where time is measured by the accumulation of the lead decay products formed at 4,567 ($\pm$1) Ma. This age is now generally accepted as "time zero" for the Solar System. The U-Pb method gives the most precise and accurate ages for these ancient objects partly because the radioactive decay constants for ^{238}U and ^{235}U are precisely known.

Once the absolute time is established using the long-lived radioactive isotopes of uranium, the sequence of events within the first few million years of the Solar System can be studied using isotopes with much shorter half lives (extinct radionuclides). These isotopes were present in the early Solar System because they had been produced in stars just prior to the beginning of the Solar System and were part of the molecular cloud that collapsed to form the Sun. Subsequently, virtually every atom of these short-lived radioactive isotopes that existed at time zero has now decayed to the daughter isotope. The isotopes used for this purpose are ^{26}Al, ^{53}Mn, ^{244}Pu, ^{182}Hf, ^{60}Fe, and ^{129}I and their corresponding decay products ^{26}Mg, ^{53}Cr, ^{136}Xe, ^{182}W, ^{60}Ni, and ^{129}Xe. The resulting sequence of events is summarized in the figure.

How old is Earth? Although the start of the Solar System is well dated at 4,567 Ma, at that time and shortly after only the pieces that would eventually come together to make Earth were present. About half or more of the planet was probably assembled by 4,550 Ma, and the Moon-forming impact, now generally thought to culminate the main phase of Earth's formation, happened at about 4,530 Ma. Earth probably continued to accumulate small amounts of material, some of them perhaps quite significant chemically, until as late as 4,450 Ma. A short episode of renewed accretion may have occurred much later, at 4,000 to 3,900 Ma.

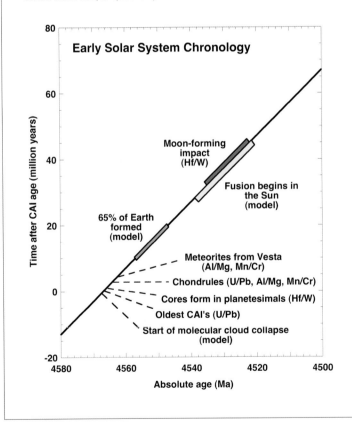

Summary of recent geochronological data and models for the sequence and timing of events in the early Solar System. SOURCE: Adapted from Halliday (2006).

BOX 1.2 Presolar Grains

On the basis of characteristically anomalous isotope ratios (Lewis et al., 1987), we now recognize and can study "presolar grains"—bits of stardust manufactured by individual stars before the birth of our Solar System that are preserved in primitive meteorites. Each of these grains contains chemical elements that were made or reprocessed by an individual star. How stars produce the heavier elements (from iron to uranium) was highlighted in *Connecting Quarks with the Cosmos* as one of the 11 major science questions for cosmology in the new century (NRC, 2003a). Geochemists will play a key role in addressing this question because the relative abundances of elements and isotopes in the different types of presolar grains provide the most specific and detailed data for checking our understanding of how chemical elements are produced in different types of stars (Zinner, 2003).

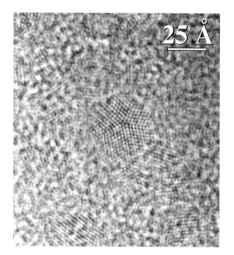

Nanodiamonds:
2 nm; ~1400 ppm

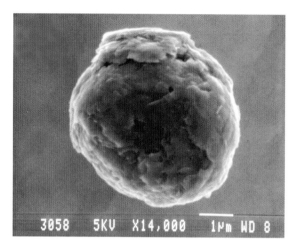

Graphite:
1 - 20 µm; ~5 ppm

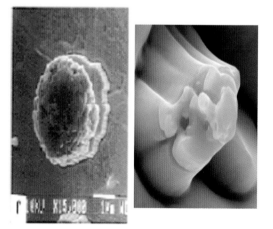

Oxides (corundum, spinel):
0.5 - 5 µm; ~100 ppb - ~1 ppm

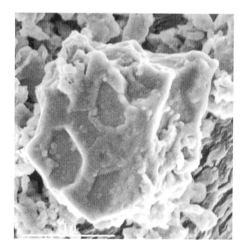

SiC:
0.3 - 20 µm; ~20 ppm

Electron microscope images of presolar grains representing materials that were manufactured by individual stars and condensed in the outflow of material marking the end of that star's life cycle. Typical sizes are given in microns (µm), and typical abundances are given in parts per million (ppm) and parts per billion (ppb) by weight. SOURCE: Nanodiamond image courtesy of Tyrone Daulton, Washington University, *Meteorite Magazine*; graphite image courtesy of Sachiko Amari, Washington University; oxide image courtesy of Larry Nittler, Carnegie Institution of Washington. Used with permission. SiC image from Bernatowicz et al. (2003), copyright 2003 by Elsevier Science and Technology Journals. Reproduced with permission.

that have little water are derived from the inner asteroid belt (inward of 2.5 AU), while the volatile element-rich meteorites, some with as much as 20 percent water as well as complex organic compounds, come from farther than 3 AU. These objects from the asteroid belt region may have been the source of Earth's water and carbon. There is also evidence that much later in the history of the Solar System—500 million to 600 million years after its formation—a large but unknown amount of rocky debris was flung into the inner Solar System, bringing a last barrage of large impacts and finishing off the major construction of the inner planets. However, it is unlikely that this "late heavy bombardment" added enough material to significantly affect Earth's overall composition.

The aspect of Earth's composition that is likely best known is the proportion of refractory elements, which form solids at the high temperatures thought to have prevailed in the inner Solar System as the terrestrial planets were forming. Included among the refractory elements are most of Earth's major components—Si, Mg, Al, and Ca. The relative amounts of refractory elements do not vary much among different classes of the benchmark chondritic meteorites, which is generally taken as a strong argument that Earth is not much different from the meteorites. For the more volatile elements, which evaporate more easily, there are wide and puzzling variations throughout the Solar System. Oxygen is one example. Si, Mg, and Fe readily combine with oxygen to form SiO_2, MgO, and FeO. On Earth almost all of the Si and Mg occur as oxides, but only about 20 percent of the Fe is combined with O; the rest is metallic Fe that resides in Earth's core. The size of the core is therefore a rough measure of the amount of oxygen that Earth has. Most meteorites have different Fe/FeO ratios, and at least two of the other terrestrial planets have a different ratio of metallic core to silicate mantle. Elements of intermediate volatility also raise important questions of chemical evolution. Potassium, for example, is relatively volatile, and estimates suggest that Earth has about 10 percent of what was available in the nebula. But exactly how much? The answer is critical because the isotope ^{40}K is radioactive and provides 20 to 40 percent of the heat produced in the early Earth. This heat plays a role in powering the convection in the mantle that drives plate tectonics (Questions 4 and 5).

The chondritic model and the Solar System's apparent ability to sort chemical elements according to their volatility have proven useful for understanding many aspects of planet formation. But our increasing ability to probe the chemical and isotopic compositions of meteorites and our planet is causing some serious rethinking of long-held models. Unanticipated compositional differences have been discovered between Earth and meteorites and between different types of meteorites. Perhaps the most striking difference is that of the isotopes of oxygen—the most abundant element on Earth (Figure 1.6). Chondritic meteorites have a peculiarly variable proportion of the isotope ^{16}O, and almost every class of meteorites has different proportions of the three oxygen isotopes. Chondritic meteorites, long thought to be the best model for the original Earth, are not like Earth with respect to oxygen isotopes. The one class of meteorites that is like Earth in this respect—enstatite chondrites—would probably be no one's first choice for Earth's main building blocks because they do not match Earth for most other ele-

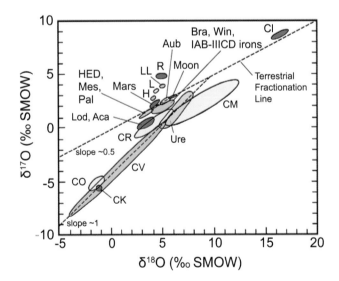

FIGURE 1.6 Representation of the range of values of oxygen isotope ratios on Earth, the Moon, Mars, and different classes of meteorites, including carbonaceous chondrites (CI, CK, CM, CO, CR, CV); ordinary chondrites (H, L, LL); other chondrite groups (R); primitive achondrites (Acapulcoite [Aca], Brachinite [Bra], Lodranite [Lod], Winonaite [Win], Ureilite [Ure]); Howardite, Eucrite, Diogenite [HED] achondrites; aubrite achondrites (Aub); stony-iron meteorites (Pallasites [Pal], Mesosiderite [Mes]); and iron meteorites (IAB-IIICD irons). SOURCE: <http://www4.nau.edu/meteorite/>. Used with permission.

ments. Moreover, it has recently been reported that the isotopes of neodymium, a lanthanide element that has proven critical for understanding planetary processes (Question 4), are also present in different amounts on Earth and chondritic meteorites (Figure 1.7), as are the isotopes of hafnium and barium.

Although we have long assumed that the isotopic compositions of the elements of the Solar System were mostly homogeneous, and measurements have borne this out in large measure, improved sensitivity is now showing small but significant differences between various planetary bodies. The O isotope differences,

for example, suggest that the nebular disk was not entirely homogeneous. While this is a problem in one sense, it is also an opportunity. If we can understand how this heterogeneity arose or was preserved, and what its structure was, we can learn more about how the materials of the nebula were sorted and gathered to produce the planets.

The Nd isotope discrepancy raises a different problem that has not yet been squarely addressed. Studies of asteroids and meteorites show that the process of accretion, whereby small chunks of rock gradually coalesce to form larger and larger bodies and eventually planets, is not one directional. When objects collide, they are almost as likely to blow each other apart as they are to coalesce. In addition, there is evidence that small accreting bodies become hot enough to melt, allowing crystals and liquid to separate. Thus, it was possible to differentiate (make heterogeneous by internal processes) smaller bodies and then blast material off them that is chemically different from the bulk object. This process would create differentiated objects that could eventually become part of the planets (or be lost into the Sun or ejected from the Solar System). In this view we cannot expect even the refractory elements to be present in exactly the same proportions everywhere, and this would have enormous implications. For example, if we relax the requirement that Earth be exactly chondritic for the elements Nd and Sm, we reach a different interpretation of the subsequent evolution of Earth's mantle and crust (Question 4). If the Hf/W ratio of Earth is not chondritic, the timing of formation of Earth's metallic core, as estimated by W isotope data, changes (see Question 2). We now know that even small bodies were able to partially melt and differentiate into core and mantle and that the mantle could potentially be removed from the core by an impact. So the timing and mechanism of formation of planetary metallic cores and the abundances of trace metals in planetary mantles have to be viewed in this context.

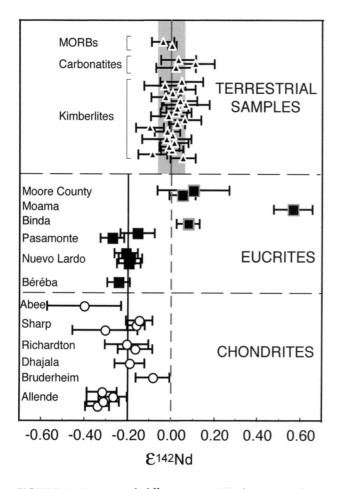

FIGURE 1.7 Reported differences in ^{142}Nd isotopic abundance between Earth, achondritic meteorites (Eucrites), and chondritic meteorites. The ε^{142}Nd value is the difference in the proportion of ^{142}Nd expressed in units of 0.01 percent. ^{142}Nd is the radioactive decay product of the short-lived isotope ^{146}Sm. The differences may reflect deep sequestration of ancient crust formed in the early Earth or differences in refractory element ratios between Earth and chondritic meteorites. SOURCE: Boyet and Carlson (2005). Reprinted with permission of the American Association for the Advancement of Science (AAAS).

Was the Moon Formed by a Giant Impact?

More is known about the Moon than any terrestrial planetary body other than Earth because of the rock samples collected by the U.S. and Soviet lunar missions between 1969 and 1976. The peculiarities of these lunar rocks—their great antiquity, their nearly complete lack

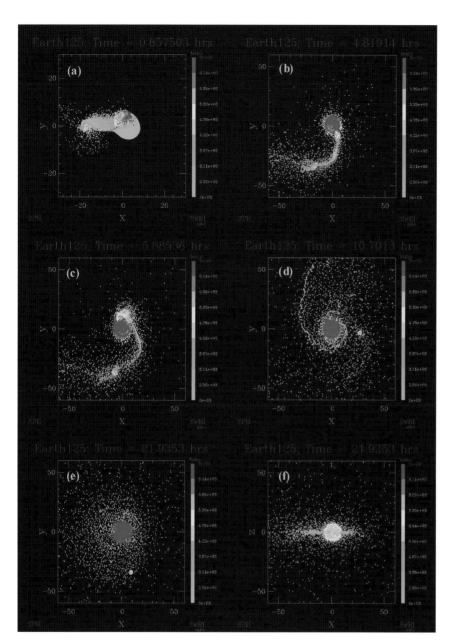

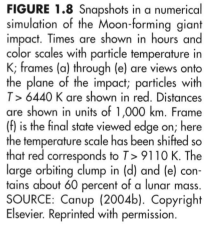

FIGURE 1.8 Snapshots in a numerical simulation of the Moon-forming giant impact. Times are shown in hours and color scales with particle temperature in K; frames (a) through (e) are views onto the plane of the impact; particles with $T > 6440$ K are shown in red. Distances are shown in units of 1,000 km. Frame (f) is the final state viewed edge on; here the temperature scale has been shifted so that red corresponds to $T > 9110$ K. The large orbiting clump in (d) and (e) contains about 60 percent of a lunar mass. SOURCE: Canup (2004b). Copyright Elsevier. Reprinted with permission.

of water and other volatile elements and compounds, and the chemical complementarity of the dark lunar basaltic lowlands and the bright highland rocks—led to enormous advances in theories of planet formation. Moon rocks provide one of the most persuasive pieces of evidence that Earth and the Moon have a common origin. The isotopic composition of oxygen varies dramatically within the Solar System (Figure 1.6) but is identical in Earth and the Moon. An important difference is the size of their metallic cores—one-third of the mass of Earth but only about 2 percent of the mass of

the Moon. Another difference is that Earth has water, as well as other volatile species and oxidized (ferric) iron; the Moon has virtually no water and all of its iron is in the reduced (ferrous) state.

Studies of lunar rocks have helped persuade many geologists that the Moon was formed when a Mars-sized object collided with the still-forming Earth about 40 million years after the formation of the Solar System. This "giant-impact" hypothesis would explain the relatively large mass of the Moon relative to Earth, the large amount of angular momentum in the Earth-

Moon system, and the chemical similarities and differences between Earth and the Moon. None of the other terrestrial planets have a moon, except for the tiny moons of Mars, which are captured asteroids.

The general features of the giant-impact hypothesis were proposed in the 1980s, but new computer models have provided a clearer picture of the requirements and results (Canup, 2004a). A "Mars-sized" object has a mass about one-tenth of Earth's, whereas the Moon has a mass about one-sixtieth of Earth's. For the hypothesis to work, the impactor must hit Earth at a low angle and at a relatively low velocity (about 10 km/s). Models indicate that most of the impactor would become mixed with and incorporated into Earth during the collision (Figure 1.8) and the cores of the two planets would coalesce at the center of gravity of the combined system. The collision would eject a disk of molten rock and vapor into orbit around the newly enlarged Earth, and a portion of that disk would coalesce into the Moon. The energy of the impact would have melted virtually the entire Earth and may have resulted in the loss of most of Earth's volatile elements (Question 2).

The impactor event, coming late in Earth's formation, would have had an enormous effect. Many of Earth's features may have been determined by the catastrophic collision, which marked the conclusion of the main phase of Earth's formation. Any internal structure that formed within Earth's mantle up to that time would probably have been destroyed, and the intense heating could have homogenized large parts of the interior. If the impact hypothesis is indeed correct, it dispels any doubt that the earliest Earth was extremely hot. The next section resumes the story of the early Earth with the aftermath of the Moon's formation.

Summary

Many lines of recent evidence have provided critical information about how and when the Solar System began and how the planets formed. Astronomical observations from increasingly powerful telescopes have added a new dimension to models of star and planet formation, as have studies of asteroids, comets, and other planets via spacecraft. There is increasing crossover between geochemical studies and astronomical observations. With improved mass spectrometric methods, new details of meteorite isotopic compositions are forcing reevaluations of the standard models for the composition of Earth and meteorites, and studies of presolar grains are sharpening our understanding of stellar evolution and nucleosynthesis. Advanced computing capabilities are enabling more realistic simulations of nebular disk evolution, the consequences of collisions between planetesimals and planetary embryos, and the internal processes of proto-planetary bodies.

But we still do not understand the composition of Earth in enough detail to make sense of its subsequent evolution. Among the most important remaining questions are when and how Earth received its volatile components, how much of these components it still contains, whether Earth is exactly the same as chondritic meteorites with respect to refractory elements, and what the absolute concentrations of heat-producing elements are inside Earth. In a broader sense we need a better idea of the processes that formed planets during the first few million years of the Solar System, how much the planets were influenced by late events (tens to hundreds of millions of years after the beginning), how the chemical composition and size of planets were determined by early Solar System processes, and the origins of the various forms of isotopic heterogeneity.

Although theory and computation are essential tools, the starting point for posing and solving outstanding questions of Solar System evolution and planet formation remains observations and measurements of planets and other extraterrestrial objects. The materials and processes of planet formation are so varied and complex, and the scales so immense, that new breakthroughs in understanding will likely continue to follow real observations made by telescopes, spacecraft, and sensitive Earth-bound analytical equipment.

QUESTION 2: WHAT HAPPENED DURING EARTH'S "DARK AGE" (THE FIRST 500 MILLION YEARS)?

Assuming that the Moon formed as the result of a giant impact, the impact would have erased the existing rock record, adding enough heat to turn Earth into a mostly molten ball, probably to the very surface of the planet. The oldest rocks yet found on Earth are about 4,000 million years old, and there are precious few of them; only about 0.0001 percent of Earth's crust is composed of rocks older than 3,600 million years (Nutman, 2006).

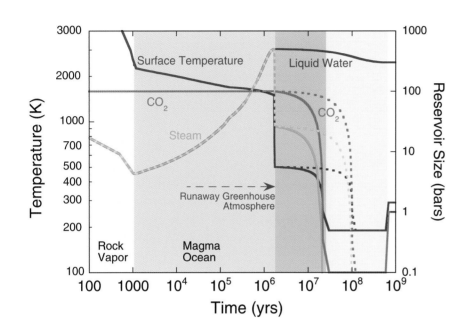

FIGURE 1.9 A speculative history of temperature, water, and CO_2 during the Hadean. The Hadean begins with the Moon-forming impact (at time = zero in this figure). For 1,000 years Earth is enveloped in hot rock vapor. After the silicate vapor rains out, the atmosphere consists mostly of CO_2. Water is gradually lost from the magma ocean and added to the atmosphere. The greenhouse effect and tidal heating maintain the magma ocean for 2 million years. When the magma surface freezes over, surface temperature drops quickly and the steam atmosphere rains out to leave a warm (~500 K) water ocean under ~100 bars of CO_2. This warm, wet Earth lasts as long as the CO_2 stays in the atmosphere. This illustration shows CO_2 being removed on timescales of 20 million years (green solid curves) or 100 million years (green dotted curves). When the CO_2 partial pressure drops below about 1 bar, the oceans freeze over (blue region of graph). After the late heavy bombardment, CO_2 is shown returning to an arbitrary level of ~1 bar, which allows the surface to be clement as required by geological data. SOURCE: Zahnle (2006). Reprinted with permission.

And most of those rocks are metamorphosed, some at very high temperature and pressure, obscuring their original form. Thus the period of time for which there is virtually no rock record on Earth extends from the time of the putative Moon-forming impact ca. 4,530 Ma to the age of the oldest rocks on Earth ca. 4,000 to 3,800 Ma. This period, about which we can discern very little from Earth itself, is called the Hadean Eon.

The name of this eon is unusually graphic, for good reason. Earth during the earliest Hadean was probably much less hospitable than even the grimmest representations of Hell. Yet somehow this inferno evolved into a place not only suitable for life but welcoming—with abundant oceans as well as dry land, an atmosphere dominated by nitrogen, and mostly comfortable temperatures. We have almost no idea how fast the surface environment evolved, how the transition took place, or when conditions became hospitable enough to support life. However, clues from Earth's oldest minerals, zircons, as well as from our Moon and other planets are

allowing a clearer picture of that early fiery (and perhaps sometimes frozen) Earth to gradually emerge.

How Did the Transition to Earth's Current Environment Occur?

Current models suggest that much of Earth's rocky mantle was melted by the Moon-forming impact and that part of it was vaporized (Stevenson, 1987; Canup, 2004a). If this was the case, liquid mantle would have been present at Earth's surface and the atmosphere would likely have been mostly rock vapor, topped by hot silicate clouds with temperatures up to 2500 K (Zahnle, 2006).[2] As Earth's surface cooled, the silicate clouds

---

[2]K represents the Kelvin temperature scale, commonly used in geology. The Kelvin scale is set so that zero degrees K is absolute zero, the temperature at which a substance has no remaining thermal energy. Zero K equals –273.15°C, and the two scales are otherwise the same, with one degree C having the same magnitude as a one-degree increment in Kelvin.

would have condensed and poured down as hot rain, perhaps at the torrential rate of about a meter a day. As the silicates rained out, gaseous compounds—especially CO_2, CO, H_2O, and H_2 but also nitrogen, the noble gases, and perhaps moderately volatile elements, such as zinc and sulfur—would become increasingly prominent.

How the transition from a hot, mostly molten mantle to something more akin to Earth's current structure happened and how fast are still matters of debate. The cooling of a "magma ocean" is a complex process, with significant uncertainties regarding the material properties of the molten and semimolten silicates, the efficiency of gas exchange between a magma ocean and the atmosphere, how much of which gases were available, and the effects of tidal heating from the Moon. We know from experiments that molten silicate would start to crystallize when the surface temperature dropped to about 1700 K and would be completely solid at about 1400 K. According to one model (Figure 1.9), the surface magma could have cooled enough for crystals to start forming after about 1,000 years and then become completely solid after about 2 million years (Zahnle, 2006). During the cooling period, most of the water and CO_2 held in solution in the magma ocean could have been vented to the atmosphere.

According to this model, solidification of the magma ocean would have taken as long as 2 million years because heat escaping from the surface would have triggered a "runaway" greenhouse state in the atmosphere, slowing the rate of heat loss (see Box 1.3). Tidal heating of Earth by the Moon would also have slowed cooling of the magma ocean (Zahnle et al., 2007). Just after its formation the Moon was much closer to Earth (perhaps half the distance) and its tidal force was much stronger than it is now. When the mantle was still completely molten, the tidal heating would have been relatively weak, but because tidal heating is concentrated wherever the mantle is solid, it would have tended to prevent the mantle from freezing.

The resultant slow cooling of the magma ocean could, in turn, have influenced the Moon's distance from Earth, which would explain why the Moon's orbit is tilted relative to Earth's orbit around the Sun (Touma and Wisdom, 1998). The relationship between the Moon's orbit and the magma ocean is somewhat complicated, but in essence the Moon could move

---

BOX 1.3 Runaway Greenhouse Effect

The runaway greenhouse effect is usually encountered as the culprit in textbook accounts of how Venus lost its water. In essence, there is an upper limit on how much thermal radiation can be emitted by an atmosphere in equilibrium with liquid water. This upper limit is called the runaway greenhouse limit, and it is about 310 W/m^2 for the modern Earth (Abe, 1993). If the planet absorbs less solar energy than the runaway greenhouse limit, all is well: the climate settles into a stable balance between photons absorbed and photons emitted. But if the planet absorbs more solar energy than the runaway greenhouse limit, the planet cannot balance its energy budget and its surface heats up. The heating continues until all the water, including clouds, has evaporated. For Earth, total evaporation of all water would leave a deep H_2O-CO_2 atmosphere over a sea of magma (Zahnle, 2006). Eventually, after some intervening photochemistry and a great deal of time, the hydrogen would be liberated from the water and lost to space. This probably happened to Venus. As our Sun brightens, this too will be Earth's fate.

For the Hadean Earth a runaway greenhouse state could theoretically coexist with a magma surface provided that sufficient water (at least a tenth of the volume of our current oceans) is present at the surface (Zahnle, 2006). The heat flow required to maintain a runaway greenhouse atmosphere (i.e., the maximum rate of cooling) would be ~150 W/m^2. Heat flow on the modern Earth is an average of 0.087 W/m^2.

---

away from Earth only as fast as it could deposit energy into Earth's mantle by tidal heating. For the Moon to lose energy to Earth's mantle efficiently, the mantle would have to be solid rather than liquid. Because solidification likely took place slowly, the Moon could drift away from Earth only at an exceedingly slow rate. This slow recession of the Moon would have allowed it to be captured into orbital resonances that gave the orbit the inclination it now has. This seemingly strange relationship between the Moon's orbit and Earth's mantle is produced by a fundamental property of planetary interiors—the dependence of viscosity on temperature—which is also critical to understanding why Earth is a geologically active planet (Questions 4, 5, and 6).

How Did Earth Develop Its Oceans and Atmosphere?

We do not know how thick the atmosphere would have been after the silicates vaporized by the Moon-forming impact condensed, largely because we are uncertain about how much gaseous material Earth contained at that stage (Question 1). The model depicted in Figure 1.9 assumes that both CO_2 and H_2O were abundant and found primarily in the atmosphere rather than dissolved in the mantle. If this assumption is correct, the initial atmosphere would have been hundreds of times thicker than the modern atmosphere, with about 100 bars of CO_2 (Zahnle, 2006). But once the surface cooled to 500 K in this model, almost all the water would rain out of the atmosphere and cover Earth with oceans, leaving the atmosphere made up mainly of CO_2. The abundant CO_2 in the atmosphere would partially dissolve in the oceans, making them acidic. At this critical juncture the level of uncertainty is redoubled. The acid oceans should then chemically attack the rocks of the seafloor, slowly turning the oceans into a soup rich in dissolved solids. If the dissolved carbon precipitated from the ocean as calcite, and if (a *big* if) the calcite was removed—for example, by dragging it down subduction zones into the mantle (see Question 5)—the CO_2 in the atmosphere would eventually be sequestered in the solid Earth, the greenhouse power of the atmosphere would gradually diminish, and Earth's surface temperature would drop.[3] Removal of CO_2 from the atmosphere might not have stopped until the CO_2 pressure was similar to today's value of about 0.0003 atmosphere. The drop in CO_2 would have lowered the surface temperature to well below freezing and covered the oceans with a global ice sheet 10 to 100 m thick. At this stage, Earth's mantle would still be exceedingly hot compared with the modern mantle, but Earth's surface would be frigid.

A key point, which links Earth's story to astrophysical models of stars, is that the Sun was 30 percent fainter in the Hadean than it is now. Given the evidence for the presence of liquid water recorded in zircons (Box 1.4), Earth would have needed abundant potent greenhouse gases to keep its surface temperature above the freezing point of water (273 K). The only good candidate greenhouse gases are CO_2 and CH_4. If CO_2 was removed by weathering and carbonate subduction, methane might work. But on today's Earth methane is made mostly by organisms, and the Hadean Earth

---

[3]Carbon sequestration by various means is discussed extensively today as a means of mitigating greenhouse warming (Questions 7 and 10).

BOX 1.4 Zircons: Earth's Oldest Minerals

Earth's oldest mineral grains are detrital zircons found in 3-billion-year-old quartzites in the Jack Hills of western Australia. Zircons are zirconium silicate crystals renowned for their durability. Because zircons incorporate uranium, their ages (a given grain may record several formation and metamorphic events) can be accurately determined from radioactive decay. Many zircons have been found that are more than 4 billion years old, and the oldest one is 4.4 billion years old (Cavosie et al., 2005). Their existence suggests there were stable continental platforms on Earth's surface in the Hadean. That such zircons have been found in only one place so far may suggest that such stable platforms were oddities rather than the rule.

The origin of zircons is also recorded in their oxygen isotopes. A mild fractionation of the oxygen isotopes suggests that many zircons formed in melts that incorporated rocks that reacted with liquid water (Mojzsis et al., 2001; Wilde et al., 2001). The zircons are silent on whether the water was 273 or 500 K, but they suggest that Earth's oceans were in place by 4.4 Ga (billion years ago). A small number of old zircons are reported to have more strongly fractionated oxygen isotopes (Mojzsis et al., 2001), which implies that the melts incorporated sediments weathered by water to make something like a granite. These data are controversial.

Zircons also incorporate hafnium, an element with a strong chemical likeness to zirconium. A deficit of radiogenic Hf in Hadean-aged zircons suggests granitic (continental type) crust was already formed at 4.5 Ga (Harrison et al., 2005).

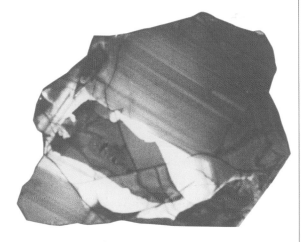

The mineral zircon is resilient to alteration and recrystallization and also contains high uranium content, which provides a means of dating individual crystals. The cathodoluminescence image of a zircon crystal shown here is 4.4 billion years old, the oldest known mineral on Earth. Zircon is the only known survivor of the Hadean period on Earth. SOURCE: Courtesy of John Valley, University of Wisconsin. Used with permission.

likely had little or no life. Without methane a mechanism is required to keep more CO_2 in the atmosphere. Could the processes that regulated atmospheric CO_2 levels within a range that kept the surface temperature well above freezing almost all the time over the past several hundred million years (Question 7) also have operated in the Hadean?

How can we test models like the one depicted in Figure 1.9? One useful observation is that the deep Earth still contains ^{3}He, a primordial gas isotope that must have been emplaced during accretion. This tells us either that the interior did not expel all its gases during a magma ocean stage (perhaps because these gases are more soluble in mantle minerals during magma genesis than previously thought; see Parman et al., 2005; Watson et al., 2007) or that the present atmosphere was added after the Moon-forming impact. Measurements of Xe isotopes suggest that Earth lost about 99 percent of its original allotment of noble gases and that it did so at least 20 million to 40 million years *after* the giant impact (Ozima and Podosek, 1999; Halliday, 2003, and references therein). To satisfy both observations, Earth must have had an early, dense atmosphere of accreted solar nebula gas (mostly H_2 and He) so that it could first gather large amounts of He and Xe and then later lose most of them. A hot and dense primordial atmosphere of nebular gas could have provided enough thermal insulation to maintain a magma ocean even before the heating of the Moon-forming event. This proto-atmosphere could have been lost when the Moon formed, or by hydrodynamic escape or ionization due to intense ultraviolet radiation from the early Sun. But since the available data are difficult to resolve with prevailing models, we are left with many uncertainties about the earliest evolution of Earth's atmosphere.

When and How Did Earth's Metallic Core Form?

Early models for the formation of Earth's core were based on a logical scenario, now known to be incorrect, that Earth first accreted into a more or less homogeneous globe (a mixture of both silicates and iron metal), then gained heat from radioactive decay of U, Th, and K. The heating gradually decreased the planet's viscosity over hundreds of millions of years, which allowed the heavy metal to sink to the center, displacing the lighter silicates toward the surface. But evidence

from meteorites, supplemented by Hf-W isotopic measurements (Figure 1.10), now clearly shows that core formation also happened in planetesimals that were much less massive than Earth (Question 1) and hence too small to have been heated from within by U, Th, and K (Kleine et al., 2002). These planetesimal cores also formed rapidly—within a few million years of the beginning of the Solar System. Hence, it is now generally accepted that core formation on Earth began when the planet was still small and accreting and that core formation probably continued for many tens of millions of years as Earth grew. The state of the core at the time of the giant impact and the influence of this

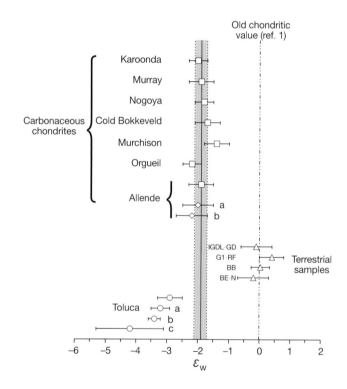

FIGURE 1.10 Comparison of the tungsten isotopic composition of Earth rocks and meteorites. Epsilon units represent deviations in the $^{182}W/^{184}W$ ratio of Earth relative to the meteorites, measured in parts per 10,000. The greater epsilon ^{182}W value of Earth relative to chondritic meteorites indicates that Earth's rocky portion formed when ^{182}Hf was still alive, which produces ^{182}W, but after most of the tungsten had been sequestered into Earth's core. Iron meteorites (open circles) have even lower epsilon values and may be representative of Earth's core composition, which is expected to be deficient in ^{182}Hf, hence ^{182}W. SOURCE: Kleine et al. (2002). Reprinted by permission from Macmillan Publishers Ltd.: *Nature*, copyright 2002.

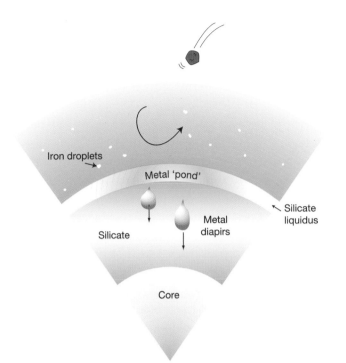

FIGURE 1.11 Possible core formation scenario during an early magma ocean in the early Earth. Small droplets of molten metal sink to the base of a magma ocean, equilibrating as they go, and pond when they reach the magma–solid rock interface. From there giant molten drops of metal (diapirs) sink through the solid but plastically deforming rocky lower mantle to reach the growing core. These diapirs do not equilibrate as they sink, so the overall pressure and temperature of metal-rock equilibration is set at the base of the magma ocean. SOURCE: Wood et al. (2006). Reprinted by permission from Macmillan Publishers Ltd.: *Nature*, copyright 2006.

event on core formation and metal-silicate differentiation remain open questions.

Clues about how the core formed come from studies of siderophile, or "metal-loving," trace elements (such as W, Pt, Os, and Pd). These elements are present in mantle rocks in the same relative proportions as in chondritic meteorites. This observation tells us that metal-silicate separation did not happen mainly at low pressure, where these elements would be strongly fractionated relative to one another. One hypothesis to explain this observation is that the metal and silicate last equilibrated chemically at the base of a magma ocean, where higher pressures may cause these elements to enter the metal in the required similar proportions (Righter et al., 1997; Figure 1.11). A competing hypothesis is that the mantle siderophile elements were added to Earth in a "late veneer" of meteoritic material after the core formed (see Palme and O'Neill, 2003, and references therein). If such a veneer was added (presumably sent in from the outer asteroid belt region as discussed under Question 1), it might also have included a substantial amount of volatile elements like water, sulfur, and carbon compounds. In this case, much of Earth's water and CO_2 could have been added long after the Moon-forming impact and might not have been present to form the blanketing atmosphere assumed in the discussion above.

Another unresolved issue is whether Earth had a primordial atmosphere at the time the core was forming. If it did, the core might still have substantial amounts of H, He, and other gases because the thick atmosphere would have kept the gases dissolved in the mantle, and if the mantle had enough of these gases, the core should have gotten its share as well. To address this issue we need to know more about the physics of separating metal from silicate. Did separation occur when both components were molten (e.g., in a magma ocean) or by percolation of more easily melted metallic liquid through solid rock? Experimental studies of how metallic melts behave when mixed with silicates are beginning to shed light on this issue (Hustoft and Kohlstedt, 2006; see Question 4).

How Did Earth's Earliest Crust Form and What Became of It?

A central question about the Hadean Earth concerns the nature of its crust and whether, in the absence of hard evidence, we can assess whether the crust had any similarity to the modern Earth's crust. Most approaches to this question begin with evidence from other planetary bodies—the Moon, Mars, and Venus—and from the oldest rocks and minerals found on Earth (see Box 1.4). The results are so far inconclusive, but the nature of the debate is rapidly changing as a result of new observations.

Earth today has two kinds of rocky crust, both of which are chemically different from the mantle (see Questions 4 and 5). Oceanic crust is relatively simple and is typically composed of solidified basaltic magma melted from the mantle. It forms by a well-understood process at midocean ridges and returns to the mantle

FIGURE 1.12 The heavily cratered light-colored areas of the lunar surface, the lunar "highlands," reflect the intense rain of meteorites that occurred in the earliest history of the Solar System. The highlands are composed of rock made mostly of a single mineral, plagioclase feldspar, which floated to the surface as the magma ocean crystallized, at about 4,500 Ma. The large, dark lunar "seas," or maria, are huge impact basins that formed mostly between 4,000 and 3,900 Ma and are evidence of a late heavy meteorite bombardment that would also have affected Earth (see Box 1.5). The lunar maria are filled with dark lava flows of basalt that formed 3,900 to 3,300 Ma. The lower crater density in the maria indicates that the meteorite flux dropped off considerably by the time the lava flows formed. SOURCE: <http://www.nasa.gov/multimedia/imagegallery/image_feature_25.html>.

by moving downward in subduction zones. Oceanic crust is thin (about 6 to 8 km), submerged under the oceans, and relatively young; its average age is about 60 million years, which is only 1.4 percent of Earth's age. The continents, which are mostly above sea level, are underlain by a different kind of crust. Continental crust is a quilt of rocks of vastly different compositions, textures, and ages and forms by multistage processes that are only partly understood. It is also thick (30 to 80 km), more silica rich than basalt, and generally old. The average age of continental rocks is about 2,000 million years, but they range from 4,000 million years

to effectively 0 million years. Like oceanic crust, continental crust appears to be "recycled" to the mantle, but at an unknown rate. The surface of average continental crust stands about 5 km higher than the surface of the average oceanic crust, so Earth's water is collected in the basins underlain by oceanic crust, and there is abundant dry land rather than a globe-encircling ocean.

Crusts are widely variable throughout the Solar System and offer no clear insight about what Earth's earliest crust was like. Samples returned by astronauts showed that the Moon's light-colored highland crust is very old (ca. 4,400 million years) and probably formed from feldspar crystals that floated to the surface after the Moon-forming impact when it was largely molten (Figure 1.12). The crust of Mars appears to be variable in age, but most is extremely old (Frey, 2006). The crust of Venus is much less well known, but a large fraction is thought to be young (Hansen, 2005; Basilevsky and Head, 2006). The crusts of the larger moons of Jupiter and Saturn seem to resemble our conceptions of the early Earth in interesting ways. Jupiter's moon Io, for

FIGURE 1.13 Image of Jupiter's moon, Io, from the National Aeronautics and Space Administration's Galileo spacecraft. Io is a volcanically active miniplanet, with young crust and no plate tectonics. SOURCE: <http://www2.jpl.nasa.gov/galileo/callisto/PIA00583.html>.

example, which is rocky and about the size of Earth's Moon, is thought to have a young crust (Figure 1.13), which it resurfaces rapidly by continuing volcanism. However, none of the rocky planets or moons have Earth's crustal resurfacing mechanism of plate tectonics (see Question 5).

One of the most obvious qualities of Earth's early crust is that it no longer exists. Why did it all disappear? The type of crust most likely to be preserved is continental crust, since virtually no oceanic crust endures for more than about 200 million years before descending into the mantle at subduction zones (Questions 4 and 5). One possibility is that the early Earth had only an oceanic-type crust and no continents. However, virtually all of the rocks preserved from the period 4,000 to 3,600 Ma are continental (3.8 Ga ophiolite in Greenland is an exception; see Furnes et al., 2007), and the only earlier materials are tiny zircon crystals that presumably also come from continental-type rocks such as granite. Isotopic evidence suggests the presence of pre-3.8 Ga continental crust, although the relative proportion varies with isotopic system (Nd isotopes suggest a greater proportion of ancient crust than Hf isotopes; Bennett, 2003). The fact that some of the oldest rocks are water-deposited sediments (3,800-million-year-old rocks from Isua, Greenland) also indicates that there was erosion and transport of sediment, which requires land standing above sea level at that time (Figure 1.14). At the average rate that Earth has been producing continental crust over the past 2 billion years, we would expect one-fifth the mass of the present continental crust to have been produced in the Hadean. However, the total volume of rocks older than 3,600 million years is very small—about 0.0001 percent of the continents. The recent observation that every Earth sample measured is enriched in ^{142}Nd compared to chondritic meteorites suggests very early formation of a crustal component enriched in incompatible elements (such as the light rare-earth elements) and its removal from the accessible portions of Earth (Boyet and Carlson, 2005). If this interpretation is correct, Earth's original crust may lie sequestered in the deep Earth today.

The uncertainties about any aspect of Hadean crust are large. Under the conditions of the Hadean Earth, which was hotter, still being hit by meteorites in its waning stages of accretion, and bearing an unknown

FIGURE 1.14 Photograph of exposures of some of Earth's oldest sedimentary rocks (about 3,700 million years), from the eastern Isua supracrustal belt in West Greenland. Metacherts (light gray) are interlayered with carbonate and calcsilicate metasediments (dark gray). SOURCE: Friend et al. (2007). Reprinted with kind permission of Springer Science and Business Media.

amount and distribution of water, we do not know whether oceanic crust production was similar to that on the modern Earth, whether plate tectonics was operating, and how efficiently continental crust was being formed and recycled. The end of the Hadean, perhaps coincidentally, corresponds to the time of the "late heavy bombardment" of the Moon's surface, which produced the large lunar impact basins that were subsequently filled with basalt lava flows (Figure 1.12; Box 1.5). Earth probably experienced this bombardment as well, but it is doubtful that such intense bombardment could cause the disappearance of a large preexisting continental crust, given that low-density ancient crust is preserved on the Moon. Rather, vigorous internal convection is more likely responsible for the demise of Earth's original crust.

Summary

The geological period called the Hadean, which extends from the time of the Moon's formation to the time when the oldest Earth rocks were formed (~4.5 to 3.9 Ga), is critical to our understanding of planetary evolution. If we are ever to fully appreciate how our

BOX 1.5 Late Heavy Bombardment

A major scientific discovery that came out of the Apollo program is that at about 3.9 Ga the Moon was pummeled by several 100-km asteroids (or comets) and by hundreds of 10-km asteroids (Wilhelms, 1987). The craters they made carved the face of the Moon. Because Earth's effective cross section is 20 times bigger than the Moon's, Earth must have been hit 20 times as often. But not only was Earth hit by a hundred 100-km asteroids, statistics imply that it was also hit by a dozen bodies bigger than any that hit the Moon. The biggest would have been comparable to Vesta or Pallas, the largest asteroids now in the asteroid belt. Whether these impacts marked the tail end of a sustained bombardment dating back to the accretion of the planets or whether they record a catastrophic event, such as a sudden influx of planetesimals to the inner Solar System due to rapid migration of the giant planets (e.g., Gomes et al., 2005), is contentious but of great importance to the Hadean environment. Examples of both possibilities are shown in the figure.

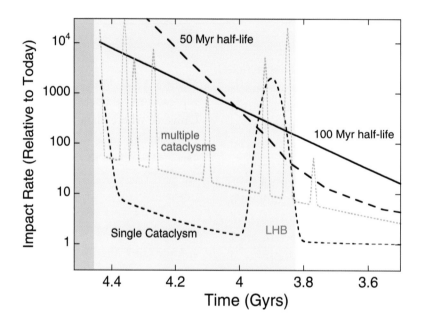

Four models of the impact rate of the first billion years of the Moon's life: a single cataclysm with a late heavy bombardment (LHB), multiple cataclysms throughout the Hadean, and sustained bombardments (denoted 50-Myr [million year] half life and 100-Myr half life). The single- and multiple-cataclysm curves are schematic representations, and the sustained bombardment curves are standard impact rates based on lunar crater counts and surface ages of the Apollo landing sites and impact basins. The 100-Myr half-life curve incorporates the age of the Imbrium impact basin and is more consistent with terrestrial and Vestan impact records than the 50-Myr half life curve, which incorporates the age of the Nectaris impact basin. SOURCE: Courtesy of Kevin Zahnle, National Aeronautics and Space Administration.

Available data offer some support for a late cataclysm, but not for the enormous hidden impacts implicit in monotonic decline. The most telling argument against a huge unseen Hadean impact flux is that it does not explain anything else in the Solar System that needs explaining. By contrast, a cataclysm (or cataclysms) fits in well with current concepts of how a solar system might evolve. All that is required is a rearrangement of the architecture of the Solar System; such rearrangements are a natural consequence of the dynamical evolution of a swarm of planets (the Moon-forming impact provides a cogent example) and are expected to occur on every timescale (Gomes et al., 2005). Before the cataclysm, impact rates would have been higher than they are today, because there were more stray bodies in the Solar System.

The impacts of the late heavy bombardment would have posed a recurrent hazard to life on Earth. Impacts by asteroids as big as Pallas or Vesta would have been big enough to boil away the oceans and leave Earth enveloped in 1500 K steam. The lunar impact record suggests that one or two of these struck Earth ca. 4.0 Ga. Conditions a few hundred meters underground would be little changed and life could have gone on (Sleep et al., 1989; Zahnle and Sleep, 1997). Later, smaller impacts may have boiled half the ocean and left the rest a scalding brine. It is this scale of event that suggests that life on Earth may have descended from organisms that either lived in hydrothermal systems or were extremely tolerant of heat and salt. It has been widely postulated that all life appears to descend from thermophilic organisms (Wiegel and Adams, 1998). Whether this means that life originated in such environments or that life survived only in such environments is debated. If the latter, the thermophilic root implies that life on Earth arose in the Hadean during the age of impacts.

planet came to be the home of complex life, we must be able to fill in this enormous gap in the geological record. At present we can construct plausible, but still highly uncertain, models for the Hadean Earth, which are based on our present understanding of planet formation (Question 1), planetary interior processes and material properties (Questions 4 and 6), and climate (Question 7). These models are informed by observations of the Moon and other planets in the Solar System, by measurements made on meteorites and the oldest rocks and minerals on Earth and the Moon, and by our geological understanding of how the modern Earth works. A critical component of understanding Hadean climate is our knowledge of atmospheric processes, but despite the advanced state of models for the modern Earth atmosphere, our understanding of radically different types of planetary atmospheres is still rudimentary.

Recent studies have raised new hope of improving our understanding of the Hadean. New information continues to be gleaned from precise measurement of the isotopic and chemical compositions of ancient zircons and their mineral inclusions. Observations of the Moon, Mars, Venus, and the moons of Jupiter and Saturn have opened new windows for visualizing the early Earth and for documenting what may have been happening in the early Solar System. Comparison of meteorites with Earth rocks has led to better models of Earth's early internal processes, including the formation of the metallic core, the implantation and loss of gaseous species from Earth's interior, and the evolution of the crust and mantle.

The future is certain to provide additional breakthroughs. Capabilities for microanalysis of geological materials are improving, and hence the amount of information that can be extracted from even the tiniest samples of old rocks and minerals is increasing rapidly. With concerted effort, it is expected that many more ancient rocks and mineral samples will be found. More precise isotopic measurements are revealing clues to early planetary processes. Planned spacecraft missions to the Moon and Mars will provide critical information about the nature of planets in the Hadean. There is even a chance that pieces of Hadean Earth rocks will be found on the surface of the Moon, sent there by impacts on Earth in the same way that pieces of the Moon and Mars have been sent here.

QUESTION 3: HOW DID LIFE BEGIN?

The origin of life stands as one of science's deepest and most challenging questions. It is a historical problem that emerged during a time with little recorded history, so it must be approached mostly through theory and experiment—imaginative efforts to re-create our planet's early conditions and establish plausible chemical routes to the emergence of life. The goal of understanding life's beginnings has attracted scientists from geology and from many overlapping disciplines, especially subfields of organic chemistry and molecular biology. In an age of planetary exploration, the origin of life is also an astrobiological issue, currently investigated on Mars, where a sedimentary record of earliest planetary history *is* preserved, and potentially across the wider stretch of Universe where planets have been detected.

Some of the most fundamental mysteries about the origin of life are geological in nature: From what materials did life originate? When, where, and in what form did life first appear? At its most basic physical level, life is a chemical phenomenon, and because it arose billions of years ago, geologists are intensely interested in creating an accurate picture of the chemical building blocks available to early life.

Top-Down and Bottom-Up Approaches

In *The Origin of Species*, Charles Darwin (1859) hypothesized that new species arise by the modification of existing ones—that the raw material of life is life. Louis Pasteur, Darwin's great Parisian contemporary, went a step further. Pasteur decisively refuted the doctrine of spontaneous generation, the long-held view that life can arise de novo from nonliving materials, declaring instead that life springs *always* from life (Pasteur, 1922-1939). These conclusions, among the most important of 19th-century science, require that forms of life developed in an unbroken pattern of descent through time, with modifications, to produce the biological diversity we see today. And indeed, students of fossils have painstakingly traced such a pattern backward for more than 3 billion years to the time of our planet's infancy (Knoll, 2003).

Before then, however, somehow and somewhere, the tree of life had to take root from nonliving precursors. Scientists have tried to identify these precursors from both the top down and the bottom up (Penny, 2005). Top-down approaches, favored by biologists, look at the complex molecular machinery of living cells for clues about simpler antecedents on the early Earth. Bottom-up approaches, pioneered by chemists, investigate the pathways by which life's chemical building blocks—the raw materials for top-down research—could have formed from simple inorganic constituents of early environments. These bottom-up approaches require the input of Earth scientists because they specify physical setting, starting materials, energy sources, and chemical catalysts. Did life originate in what Darwin envisaged as a "warm little pond," perhaps a tidal pool repeatedly dried and refreshed? Or might life be rooted among hydrothermal vents? Could life's origins even lie beyond Earth? Experimental approaches to prebiotic chemistry must be framed in terms of environments likely to have formed life's incubators, and only Earth scientists can inform us about the physical and chemical characteristics of these settings.

A Search for Clues in the Laboratory

We have understood for more than half a century that modern laboratory experiments can shed light on our search for life's beginning. In the classic Miller-Urey experiment, Stanley Miller (1953) ran an electric spark through a mixture of water vapor, ammonia, methane, and hydrogen gas, generating a complex array of organic molecules, including amino acids, the building blocks of proteins (Figure 1.15). Intermediate products in amino acid synthesis included formaldehyde (from which sugars can be synthesized) and hydrogen cyanide (the starting material for abiotic synthesis of the bases that specify information in nucleic acids).

In this experiment the spark serves as a proxy for lightning in the early atmosphere. The gas mixture approximates one hypothesis for atmospheric composition. As it turns out, the success of Miller-Urey and other experiments in prebiotic chemistry depends critically on the relative amounts of gases present in the early atmosphere and oceans. The Miller-Urey mechanism requires more hydrogen than carbon (Miller and Schlesinger, 1984), and Miller chose his starting mixture to approximate the prebiotic atmosphere as envisioned by his mentor Harold Urey. But since then most atmospheric scientists have adjusted the model to

environments that have less hydrogen and therefore are less strongly reducing (Kasting and Catling, 2003). In contrast, Tian et al. (2005) have argued that less hydrogen escaped to space from the early atmosphere than was previously assumed, which implies that while most carbon in the primitive atmosphere was in the form of carbon dioxide, hydrogen gas was also available for organic synthesis, with energy added by lightning. Impacts by iron-rich meteorites might also have transient enrichment in compounds such as carbon monoxide that would have facilitated the synthesis of biologically interesting organic compounds (Kasting, 1990).

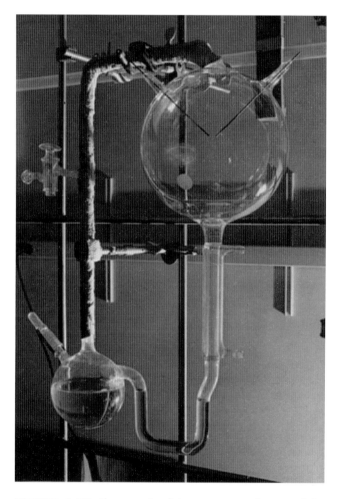

FIGURE 1.15 Photograph of the experimental setup of the famous Miller-Urey experiment. An electric spark passes through a chamber containing hydrogen gas, ammonia, methane, and water vapor; as the product of the resulting chemical reaction cools, water condenses, carrying organic molecules to the flask at the bottom of the apparatus, where they can be sampled and analyzed. SOURCE: Bada and Lazcano (2003). Reprinted with permission from AAAS.

The availability of gases such as hydrogen and carbon monoxide in Earth's early atmosphere is currently the subject of vigorous debate among Earth scientists, and its outcome will determine how we think about environmental chemistry and the origin of life during Earth's early development. Whether or not amino acids and other organic molecules were widespread on the early Earth, they existed in some parts of the early Solar System and reached our planet in the form of carbonaceous chondrites. These meteorites contain significant abundances of biologically interesting compounds, as do some interstellar clouds.

How Did Life Arise?

Earth scientists are trying to answer this question by combining field and laboratory studies of the planet's oldest sedimentary rocks, laboratory simulations, and geochemical theory to define the environmental conditions most likely to have nourished early life. A central question, for example, is what combination of the basic conditions—nitrogen and phosphate availability, electrochemical and acid-base qualities of the environment, and abundances of trace metals and minerals—were the most life enhancing? The challenge is to identify and quantify every one of these conditions to actually estimate the probability of forming life under primitive Earth conditions. Because those conditions are today poorly preserved or absent, geologists must adapt tools of many kinds to infer how life began.

Essential to our understanding of how life emerged from prebiotic chemicals is accurate knowledge of the kinds of catalysts present in the environment. A catalyst is a substance that increases the speed of a chemical reaction, often dramatically. In every cell the complex and coordinated chemical reactions that support life require the action of catalysts, usually enzymatic proteins. Many prebiotic reactions require catalysts as well, not only to support energy-yielding reactions but also to permit the synthesis of the long-chain molecules such as nucleic acids and proteins that make up living systems. Some of the most essential catalysts used in experimental approaches to prebiotic chemistry are metal ions, which coordinate chemical reactions in developing metabolism, and mineral surfaces, which provide templates and catalysis in synthesizing biopolymers.

The idea that metal ions, dissolved in early lakes and oceans, might have catalyzed prebiotic chemical reactions follows closely from our knowledge of biochemistry. Biological catalysts commonly depend on the action of a cofactor that contains a metal at its functional heart. For example, a magnesium ion occupies the center of the chlorophyll molecules that trap light energy and drive photosynthesis. Similarly, an iron atom lies at the center of hemoglobin, the molecule that transports oxygen in mammalian respiration. A wide diversity of metals act as important catalysts for biological reactions, especially iron, manganese, magnesium, zinc, copper, cobalt, nickel, and iron-sulfur clusters. Understanding the roles these metals might have played in prebiotic chemistry is a geological question whose answer depends on how the metals were distributed in primitive Earth environments. To find such answers we need integrated data about (1) early crustal differentiation and magma generation (see Question 2), (2) the low-temperature chemistry of weathering, (3) hydrothermal reactions in ancient seafloors, and (4) oxidation-reduction (redox) conditions in early environments. Once we understand these conditions, experiments in prebiotic chemistry can graduate from artificial media doped with single metal ions to complex ionic mixtures informed by Earth science.

The same is true for mineral surfaces, long recognized as potentially important catalysts of prebiotic chemical reactions (Schoonen et al., 2004; Figure 1.16). Clay minerals, for example, have been shown to catalyze the assembly of lipid micelles into vesicles—tiny spheroids that could have governed prebiotic-phase separation on the early Earth (Hanczyc et al., 2003). Clay minerals also catalyze the linkage of nucleotides to form nucleic-acid-like polymers (Orgel, 2004), and pentose sugars (including the biologically important ribose) can be stabilized in the presence of calcium borate minerals (Ricardo et al., 2004). A role in prebiotic chemistry for iron sulfide minerals has been suggested as well, most prominently in Wächtershäuser's chemically explicit theory of biogenesis around hydrothermal vents (Wächtershäuser, 1988; see Hazen, 2005, for a discussion of recent experimental tests). Continuing advances will require new experiments based on realistic mineral catalysts, as well as constrained theory, experiments, and observations from Earth science (Schoonen et al., 2004). In particular, we need to understand how

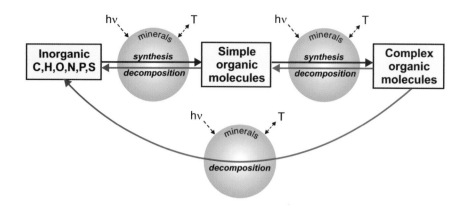
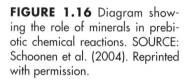

FIGURE 1.16 Diagram showing the role of minerals in prebiotic chemical reactions. SOURCE: Schoonen et al. (2004). Reprinted with permission.

chemical reactions between water and the early crust shaped the chemistry of early environments.

When Did Life Arise?

A second important question flows from the first: When did life arise on our planet? Paleontologists and biogeochemists have long agreed that the origin of life preceded the deposition of minimally metamorphosed sedimentary rock deposited 3,500 to 3,400 Ma. Tiny fossils preserved in sedimentary rocks document microbial diversity in rocks deposited long before animals evolved, and stromatolites—sedimentary structures formed by the interaction of microbial communities and the physical processes of sedimentation—provide independent evidence of widespread microbial life on the early Earth (Knoll, 2003; Figure 1.17). Because bio-

logical processes such as photosynthesis tend to enrich organic molecules in the lighter stable form of carbon (^{12}C) relative to its heavier forms, we can estimate when carbon began to be trapped by photosynthetic microorganisms. Carbon isotopic abundances in 3,500 million year old sedimentary rocks are similar to those found in much younger deposits, suggesting that a biological carbon cycle was established early in our planet's history. Indeed, highly metamorphosed rocks that are nearly 3,800 million years old contain carbon isotopic abundances suggestive of a still older carbon cycle. It has further been proposed that the high concentrations of organic matter in some of the earliest known shales require primary production by photosynthetic organisms (Sleep and Bird, 2007). In light of these observations, the close molecular similarity of all known species strongly suggests that all living organisms are descended from a common ancestor that lived nearly 4 billion years ago.

Did Life Originate More Than Once?

We cannot tell how many times life arose. Life may have originated many times on the young Earth, with the ancestor of present life persisting by good luck (chance survival of primordial mass extinctions) or good genes (outcompeting other early life forms). But experiments can help us understand whether there is more than one route to life. There is no reason these routes must all be terrestrial, and some scientists have speculated that terrestrial life was seeded from afar, most likely from Mars (Weiss et al., 2000). A mechanism certainly exists—several lines of evidence show that Earth receives a continuing stream of meteorites ejected to space from Mars by meteor impact and that

FIGURE 1.17 2.76 billion year old stromatolite in Pilbara, Australia. SOURCE: Ohmoto et al. (2005). Reprinted with permission.

some of these meteors could have delivered microbial cargo to Earth. The obvious test is to learn by exploration whether Mars was ever a biological planet. At present we do not know, but exploration of ancient sedimentary rocks on Mars, guided by our geological and paleobiological experiences on Earth, may provide an answer. From orbital observations and the in situ exploration by the Mars rovers *Spirit* and *Opportunity*, we know that Mars—unlike Earth—preserves a sedimentary record of surface environments from its first 500 million years (e.g., Squyres et al., 2004). Thus, Martian rocks might preserve a record of prebiotic chemistry, or even nascent life, if such records ever formed. Many scientists have attempted to estimate the odds that life can emerge as a lucky accident, whether on a planet or elsewhere where environmental conditions are favorable. Experiments in prebiotic chemistry will nudge us toward better answers, but what the question really requires is a second example of a living system.

In recent years, however, skeptics, stimulated in part by controversial claims about biological signatures in a Martian meteorite, have challenged the conventional wisdom that terrestrial life arose on Earth prior to 3,500 Ma. Explanations that do not involve biology have been proposed for micron-scale carbon-bearing structures previously interpreted as Earth's oldest microfossils, for stromatolites, and for carbon isotopic abundances in carbonate minerals and organic matter (e.g., Brasier et al., 2005, 2006). Vigorous defenses of biological interpretations have been mounted (e.g., Schopf et al., 2002; Allwood et al., 2006; Schopf, 2006). At present the weight of evidence favors the hypothesis that life existed 3,500 Ma, and likely existed back at least 3,800 Ma, but much remains to be learned about the nature of early ecosystems. Only careful mapping and stratigraphic analysis will tell us whether our planet preserves an earlier record of its biological (or prebiological) history, and only innovative biogeochemical analyses set in the context of well-corroborated microbial phylogeny will resolve uncertainties about the antiquity and nature of early microorganisms.

What Is Life—and What Is Not Life?

In one way, at least, biologists have it easy: they can evaluate whether a structure is living by testing for evidence of metabolic activity. Does it breathe? Does it eat? Can it move against gravity? Paleontologists have a more difficult task, necessarily judging biogenicity by shape, distribution, and chemistry. No sensible person would doubt that dinosaur skulls excavated from Cretaceous sandstones constitute definitive evidence of ancient life; no known physical processes can produce the complexities of a skull in the absence of biology. Similarly, the preservation of cholestane (the geologically preservable form of cholesterol) in a Jurassic oil tells us that life existed when the oil deposit formed because cholestane does not form abiologically. The problem gets harder when we go backward in time beyond the first appearance of animals ca. 580 Ma. Some microfossils have complicated shapes clearly related to living organisms (Figure 1.18a, b), and an unambiguous record of microfossils goes back some 2,500 million years. Older candidate fossils, however, tend to be poorly preserved and have simple shapes. The tiny spheroid structure in Figure 1.18c is about 3,500 million years old and is made of carbon. It is hard to be sure this is a fossil because such simple structures might well form from physical processes.

The same uncertainties confound investigations of larger scale features of sedimentary rocks that may have been imported by organisms, as well as molecular or isotopic features of ancient organic matter that might reflect biological processes. Stromatolites, for example, are commonly interpreted as the sedimentary products of sediment accretion on ancient lake bottoms and seafloors. Stromatolites formed by trapping, binding, and cementing sediment particles have textures not easily mimicked by purely physical processes, so they provide reliable evidence for life in rocks more than 3,000 million years old (Figure 1.19a). Other stromatolites form by mineral precipitation, however, especially in the oldest sedimentary accumulations, and it is difficult to know what role, if any, life played in their accretion (Figure 1.19b).

The challenge of identifying the geological products of life becomes even more difficult when applied to ancient rocks of Mars or other planets. We have no confidence that the diversity of life on Earth exhausts all possibilities for living systems. Thus, the guiding question of paleo- and geobiological exploration of the Solar System is whether a structure (molecular, microscopic texture, or stromatolite) found during planetary exploration can be explained adequately in terms of

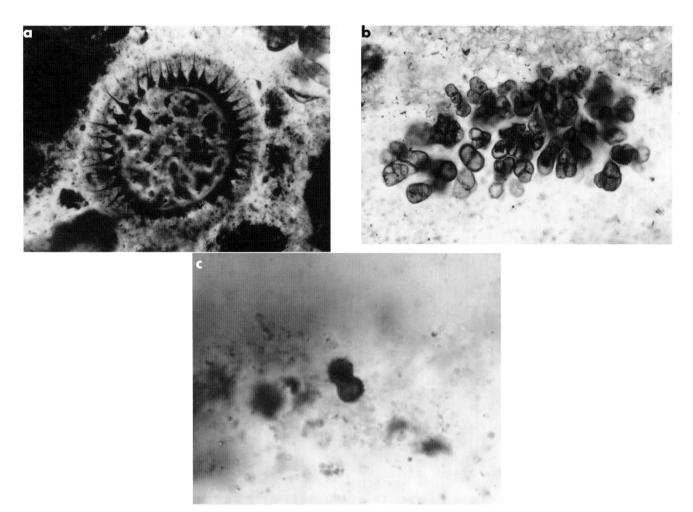

FIGURE 1.18 (a) Fossil of a eukaryotic microorganism preserved in ca. 580 Ma phosphorite from the Doushantuo Formation, China. The fossil is 250 microns across. (b) Branching cyanobacterium preserved in ca. 800 Ma chert from the Upper Eleonore Bay Group, Greenland. The fossil is 500 microns from left to right. (c) Paired 4-micron-wide carbonaceous spheroids in ca. 3,500 Ma cherts from the Onverwacht Group, South Africa. Are these fossils? SOURCE: Courtesy of Andrew Knoll, Harvard University.

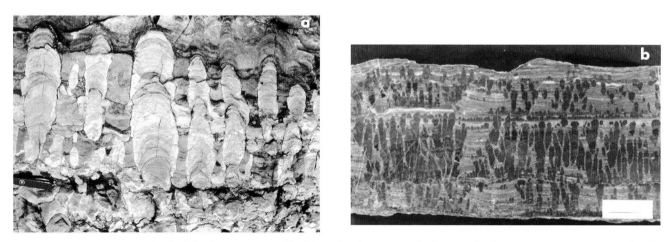

FIGURE 1.19 (a) Stromatolite built by the trapping and building of sediment particles by microbial communities—1,500 Ma Bil'yakh Group, Siberia. SOURCE: Courtesy of Andrew Knoll, Harvard University. (b) Stromatolites built of seafloor precipitate structures that are composed of calcium carbonate crystals without any obvious templating influence of microorganisms—1,900 Ma Rocknest Formation, Canada. SOURCE: Courtesy of John Grotzinger, Caltech. Used with permission.

known physical processes. Some molecular and morphological structures form only by biological processes (cholesterol, dinosaur skulls), while others clearly relate to physical processes (large quartz crystals, for example), and still others exist in a zone of overlap (2-micron spheres, amino acids). We can never eliminate the zone of overlap, but better understanding of the products of both biological and physical processes will better equip us to pursue questions of life's antiquity on Earth and its distribution through the Solar System.

Is There Life Beyond Earth?

Our understanding of our own origins remains sketchy, but it is expanding at an accelerating pace. Thanks to contributions from many fields and approaches, scientists are better prepared to approach a truly tantalizing question: Are we alone, or has life also evolved elsewhere? If life exists elsewhere, what forms does it take? With continuing planetary exploration, Earth scientists will be able to establish with greater certainty whether life could have originated elsewhere in our Solar System—and even whether organisms could have become established on Earth by meteoritic transfer from another planet. Thanks to discoveries of the National Aeronautics and Space Administration's rover *Opportunity*, we now know that around the time life took root on Earth, at least regional environments on Mars' surface were episodically wet (Knoll et al., 2005). But they were also oxidizing and strongly acidic—serious obstacles to many of the prebiotic chemical pathways thought to have been important on Earth. Was early Mars arid, oxidizing, and acidic globally or just regionally, and when were such environments established?

Clay minerals in some of Mars' oldest terrains may signal that early in its history our neighbor was relatively wet but less acidic (Bibring et al., 2006). Also, carbonate and sulfide minerals precipitated from fluids flowing through crustal fractures document at least transient subterranean environments neither strongly acidic nor oxidizing (McKay et al., 1996). Only further exploration, with Earth and planetary scientists working in partnership, will establish whether life on Earth is unique in our Solar System or merely uniquely successful.

Summary

While synthetic organic chemistry and molecular biology will continue to provide the experimental basis for probing life's origins, Earth scientists will increasingly specify the conditions under which laboratory experiments are run. Stratigraphers, paleontologists, biogeochemists, and geochronologists can provide sharper constraints on when life arose and the metabolic character of early organisms. Geochemists focused on both crustal differentiation and low-temperature reactions can build an improved sense of redox conditions, weathering reactions, and metal abundances on the early Earth. Modelers can use new data to provide more sophisticated hypotheses about how our planetary surface operated in its infancy, setting the stage for the intercalation of biological processes into the Earth system. And planetary scientists, now exploring Mars and other bodies at a resolution previously limited to Earth, can provide comparable insights about environmental (and, at least potentially, biological) evolution on other planets.

2

Earth's Interior

As planets age they slowly evolve as the heat trapped and generated in the interior is transported to the surface. The internal planetary processes that move this heat—including volcanism and convection—have a huge influence on the nature of planetary surfaces. Yet the vast interior is inaccessible to direct study and must be understood with geophysical observations, experimental studies of materials under deep-Earth conditions, and theoretical models. For over a century, seismic wave, geomagnetic, and gravity measurements made at the surface have been improving our characterization of Earth's internal structure. Experimental and theoretical determinations of material properties at high temperatures and pressures and numerical modeling of mantle and core heat transport and convection over very long timescales also play key roles in studies of internal dynamics. However, despite continuing advances, we still cannot uniquely describe Earth's mantle structure or explain in any detail how the core and mantle work, why Earth differs from other planets, or how it may change in the future.

The three questions included in this chapter describe scientific challenges for understanding Earth's evolution and internal dynamics. Question 4 addresses deep-Earth dynamical processes, from the inner metallic core at the center of Earth to the convecting mantle to the volcanoes at the surface. Question 5 focuses on the near-surface features of Earth—old continents, young ocean basins, and plate tectonics—that make Earth unique among Solar System planets and that also seem inextricably linked to the presence of water and the preservation of life-sustaining conditions. Ques-

tion 6 deals with Earth materials properties, which control many of the internal processes discussed in this chapter.

QUESTION 4: HOW DOES EARTH'S INTERIOR WORK, AND HOW DOES IT AFFECT THE SURFACE?

The previous chapter discussed evidence that Earth and the Moon, and by extension the other terrestrial planets, started out with high internal temperatures about 4.5 billion years ago. Once the planetary accretion process tails off, the planets cool, first through a period of active geological processes and ultimately to a state of geological quiescence. When the planet is geologically active, evidence of that activity is reflected in the nature of its surface and atmosphere and perhaps the existence of a magnetic field. After the interior cools and its viscosity increases sufficiently, geological activity grinds to a halt, and the planet's surface stops regenerating. Thereafter, only external processes, such as bombardment with asteroids, further modify the surface.

Some planetary bodies, like the Moon, cooled quickly and have been geologically inactive for billions of years. Despite rapid cooling after the Moon-forming impact (Questions 1 and 2), Earth produced and retained enough heat to power geological activity until the present, and it is likely to do so for several billion more years. However, both the amount of Earth's cooling and resulting changes in the internal dynamics and surface environment are still poorly known. Although we know that heat is transported by mantle convection, we do not yet have the capability to exactly describe these convective patterns, calculate with confidence how different they were in the past, or predict how they will change in the future. Resolving the critical questions about planetary evolution will require much more advanced knowledge of planetary materials and how they affect convection (Question 6), better constraints from seismology on the present configuration of mantle flow at both large and small scales, and significant advances in mathematical modeling of convection that is driven by both temperature and chemical variations.

Convection and Heat Flow

About 43 TW (10^{12} J/s) of heat flows from Earth's interior through its surface at present, based on global heat flow measurements and thermal models for cooling oceanic lithosphere. Sources of this surface heat flow include the slow cooling of the mantle and core over the history of the planet; heating produced by radioactive decay of U, Th, and K; and minor sources such as tidal heating. The exact contribution of each to the planet's heat flow is uncertain. For example, we do not know how much U, Th, and K are contained inside Earth and how these elements are distributed (McDonough, 2007). These elements are more effective at keeping Earth hot if they are located deep within the mantle, or even to some degree in the core, rather than near the surface. As a result of these uncertainties, we cannot yet answer the simple question: How fast is Earth cooling?

The primary mechanism for transporting heat within Earth's interior is convection. It was once believed that mantle convection was impossible because the mantle was demonstrably solid. But much like a glacier, the mantle can behave like both a brittle solid and a liquid: it fractures when deformed rapidly but flows on long timescales. We now know that both the mantle and the outer core circulate in a complex pattern of large- and small-scale flows. In the molten outer core, which has very low viscosity (some estimates suggest a value similar to that of liquid mercury), convection is rapid. Hot liquid metal circulates up to the top of the core where it loses heat to the base of the

mantle and then sinks again in a turbulent pattern that is affected by rotation and the magnetic field the flow generates. By contrast, mantle motions are ponderous. Typical velocities are about 5 cm/yr (based on geodetic, magnetic, seismic, and geological measurements), and at this rate the nominal "round-trip" journey of a mantle wide convection cell—across the surface for 5,000 km, down 2,900 km to the bottom of the mantle, and back to the surface again—would take about 300 million years. This rate of travel is consistent with simple thermal convection models that treat the mantle as if it were a liquid with a viscosity (estimated from postglacial rebound rates) of about 10^{21} Pa-s. The configuration of convection in Earth's mantle provides the primary control on how Earth cools, mainly because the mantle makes up roughly two-thirds of Earth's mass and 85 percent of its volume (Figure 2.1).

Mantle motions carry hot material from deep inside Earth toward the surface, where heat is lost to the atmosphere and ultimately to space, and also carry cold surface rocks down to great depths. Unresolved issues concerning mantle convection arise from uncertainties about material properties at high pressures and temperatures. Experiments and field evidence show that mantle rock becomes soft enough to flow over geological time periods at depths of just 30 to 60 km, where the temperature surpasses 700°C and pressure reaches several thousand atmospheres. At higher temperature—above 1200°C—the viscosity of mantle rock is low enough that it behaves much like a thick liquid; almost all of the mantle is hotter than 1200°C. Mantle viscosity exerts the primary control on the form of convection and the efficiency at which heat is moved toward Earth's surface. However, other factors also are important. For example, viscous dissipation associated with deformation of stiff lithospheric plates at subduction zones strongly affects the form of convection and the relationship between convective vigor and surface heat flow. The largest uncertainties are for the lower mantle. Seismological data suggest that the flow pattern there is complex. Other observations suggest that viscosity increases in the lower mantle, and numerical models indicate that flow velocities in the lower mantle may be much slower than plate velocities such that the overturn time is a billion years or more (Kellogg et al., 1999; Ren et al., 2007).

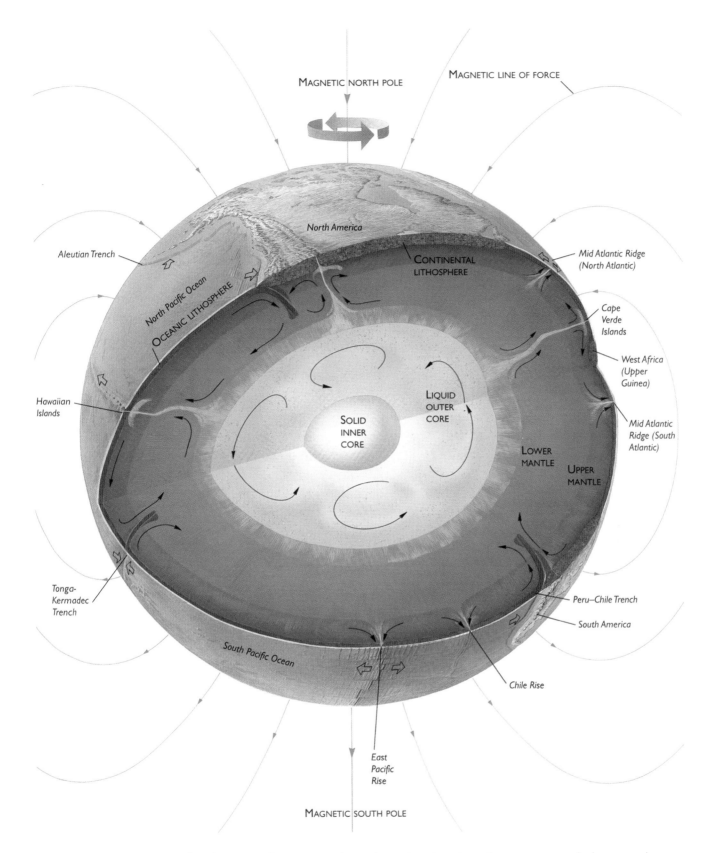

FIGURE 2.1 Cutaway view of Earth's interior showing major layers (oceanic and continental crust, upper mantle, lower mantle, outer core, inner core) and features (mantle plumes, subduction zones, midocean ridges, convection currents, magnetic field). SOURCE: Lamb and Sington (1998). Copyright 1998 Princeton University Press. Reprinted with permission.

How Are Mantle Convection and Earth's Thermal Evolution Related?

We know that mantle convection is driven by the heat of Earth's interior, but what controls Earth's temperature? The current understanding is that the mantle itself acts as Earth's primary "temperature regulator," and its actions depend on the atomic-scale properties of mantle minerals that determine viscosity. The effective viscosity of the mantle depends on the rate at which the mineral grains can deform in response to an applied stress, which in turn is strongly dependent on temperature. Laboratory data indicate that for a given stress a 100°C temperature increase lowers the viscosity by about a factor of 10. Consequently, if Earth were to heat up, it would convect more vigorously and lose heat faster. As heat is lost, temperature drops and convection slows, decreasing the rate of heat loss. This temperature-viscosity feedback should keep Earth's internal temperature well regulated. The temperature at which the thermostat is most likely to be set is just below the melting point of mantle rock because there is an even faster decrease in viscosity with temperature once the mantle begins to melt.

The temperature-viscosity feedback model is useful but it implies a steady system that undergoes only slow changes over long periods of time. This implication is at odds with much of what we know and suspect about mantle materials and geological history. For example, the continents, which are an end product of Earth's evolution, show evidence of rapid growth spurts (Question 5), which may or may not be associated with accelerated plate tectonics (Hoffman and Bowring, 1984). The seafloor of the western Pacific Ocean contains enormous volcanic mountain ranges, which suggests that the Cretaceous Period (65 to 150 Ma) was a time of exceptionally intense volcanic activity and possibly also fast seafloor spreading (Engebretson et al., 1992). We also know that the Cretaceous was a period of exceptional global warmth and high sea level (Question 7) and stability of Earth's magnetic field. These observations as well as theoretical considerations raise the question of whether Earth's thermal evolution and internal processes are adequately described by our (quasi-) steady state models or whether the evolution has been unsteady and punctuated by catastrophic reconfigurations. Thus, even though we understand the most basic features of mantle convection, our level of

understanding is insufficient to explain many of the most important geological and geochemical features of our planet.

We are even further from understanding the internal evolution of other rocky bodies of our Solar System, where we have fewer data, and interactions between thermal evolution and orbital evolution provide additional complications (see Box 2.1). Earth (and possibly Venus) has apparently maintained a high enough internal temperature to ensure continued geological activity. However, on smaller planetary bodies, geological surface activity has either long since stopped (Moon) or slowed greatly (Mars). It is believed that the mantles of other terrestrial planets should function in the same way as Earth's, unless there are different amounts of radioactive elements or different amounts of water dissolved in the mantle minerals. The addition of tiny amounts of water to mantle minerals would lower both the viscosity of the mantle and the melting temperature (Question 6) and may prolong a planet's geologically active life.

What Do Mantle Plumes Tell Us About Convection and Heat Transport?

The viscosity of Earth's mantle is sufficiently low and sensitive to temperature that convection can include complex small-scale currents. Evidence of this small-scale convection is provided by hot spots—large clusters of volcanoes, the most active of which are in Hawaii, Iceland, the Galapagos Islands, Yellowstone, and Réunion (Indian Ocean). Hot spots are usually explained as the surface outpourings of magma formed in mantle plumes, which are cylindrical upwellings of hot (and hence low viscosity) rock that are thought to form near the base of the mantle and rise to the surface at rates much faster than plate velocities (Figure 2.2). Mantle plumes should form as a consequence of heat entering the bottom of the mantle from the much hotter outer core.

Mantle plumes may also be responsible for large igneous provinces, which are vast basalt lava plateaus on continents and the ocean floor. The best current explanation is that they form when the bulbous top of a new plume approaches Earth's surface (Figure 2.2), then spreads out and causes widespread melting (Ernst et al., 2005). These large, rapid lava outpourings may have caused major perturbations to Earth's climate (Ques-

---

BOX 2.1 Planetary Comparisons

Our understanding of our planetary neighbors has advanced substantially over the last several decades through spacecraft exploration and analysis of lunar samples and meteorites from Mars and the Moon. The other terrestrial planetary bodies (Venus, Mars, Mercury, and the Moon) formed by the same processes as Earth (see Question 1) and are governed by the same physical and chemical laws and processes. Nevertheless, each has taken a distinct evolutionary track, deepening the questions we pose for how Earth works the way it does.

Venus, at 0.8 Earth masses, is sometimes called Earth's "sister planet," but its massive carbon dioxide atmosphere (90-bar surface pressure) and global cloud cover have led to a runaway greenhouse, a surface temperature of 470°C, and the loss of most water from the atmosphere. Venus also lacks Earth-like plate tectonics, but the planet has been subjected to resurfacing—probably by some form of lithospheric recycling not understood—at least once and perhaps multiple times. The density of impact craters indicates that the surface has an average age between several hundred million years and 1 billion years. There are mountain belts and pervasively deformed plateaus, both of which are stratigraphically older than the widespread volcanic plains, known to be basaltic from spacecraft lander measurements. Unlike Earth, Venus has no detectable internal magnetic field. A strong correlation of long-wavelength gravity and topography in the plains is the signature of ongoing mantle convection. Rifting and volcanism have occurred more recently than the average surface age, and the planet is likely to be volcanically and tectonically active at present.

Mars, at 0.1 Earth masses, evolved more rapidly than Earth or Venus. Isotopic evidence from Martian meteorites indicates that Mars formed its core, mantle, and most of its crust within a few tens of millions of years after the beginning of Solar System formation, probably without any plate tectonics era. Large segments of the most ancient crust on Mars are strongly magnetized, relics of a core dynamo that began early in Martian history but probably died out after several hundred million years. The Martian surface has seen a mix of plains volcanism and more focused magmatism in regional centers, dominated by the Tharsis volcanic province, largely constructed before 4 Ga (billion years ago). Fluvial landforms, widespread chemical alteration, and sedimentary deposits visited by surface rovers all indicate that water was an important agent of geological change early in Martian history. At about 4 Ga, Mars lost its global magnetic field, its carbon dioxide atmosphere was substantially thinned by solar wind stripping, the climate cooled, and water lost its dominant role in surface change. Martian volcanism continued at generally declining rates, and the planet may still be active at low levels today.

The Moon and Mercury, at 0.01 and 0.05 Earth masses, respectively, have heavily cratered surfaces and only extremely tenuous atmospheres, but their similarities end there. The Moon began largely molten, presumably the result of the accumulation of hot ejecta from a giant impact on the early Earth. Cooling and solidification of the resulting magma ocean led to formation of the crust and the mantle source regions of later volcanic lavas. Those lavas erupted to partially fill the lunar maria, mostly on the lunar nearside at 3 to 4 Ga, but there are also isolated younger volcanic deposits. The Moon may have a small iron-rich core, but if so it is no more than a few percent by mass. Lunar rocks from 3 to 4 Ga are magnetized, but whether the magnetizing field was a central core dynamo or transient field generated during surface impacts is an open question. The Moon is seismically active at low levels today. Shallow moonquakes are probably the signature of interior cooling, whereas deep moonquakes occur in clusters and appear to be triggered by tidal stresses.

Mercury, in contrast, has such a high bulk density that its iron-rich core comprises at least 60 percent of the planet's mass. Mercury has a global magnetic field, dipolar like that of Earth, and the outer core is known to be molten on the basis of the amplitude of the planet's libration forced by solar torques as Mercury progresses along its elliptical orbit. The planet has an ancient, heavily cratered crust, as well as somewhat younger plains units that may be volcanic in origin. The surface composition is poorly known, but Earth-based measurements indicate that surface silicate materials have little or no ferrous iron. The dominant tectonic landforms on Mercury are high-relief lobate scarps, the surface expressions of large-offset thrust faults. Because of the extensive distribution and apparently random orientation of these features, the lobate scarps have been interpreted to record an extended period of global contraction, the result of some combination of interior cooling and solidification of an inner core.

---

tion 7) and perhaps even major extinctions (Question 8). Other indications of plumes include broad bulges in the ocean floor, such as those around Hawaii, and the tremendously excessive amount of lava produced at Iceland in comparison to other places along the Mid-Atlantic ridge.

Although there is good geological evidence that mantle plumes exist, seismological evidence for the existence of narrow, hot, cylindrical upwellings in the lower mantle is only equivocal. Some cylindrical regions of low velocity appear to extend downward to 200 to 600 km, while others seem to extend almost to the core-mantle boundary. However, there is abundant evidence for much larger, domical or irregularly shaped low-velocity features in the lower mantle that are sometimes called superplumes (Figure 2.3). Does this mean that thermal plumes do not exist in the lower mantle or that the seismic resolution is still too low to make them out? Seismic data suggest that the large low-seismic-velocity regions near the base of the mantle are anomalously dense, which is contrary to expectations for buoyant thermal upwellings (Ishii and

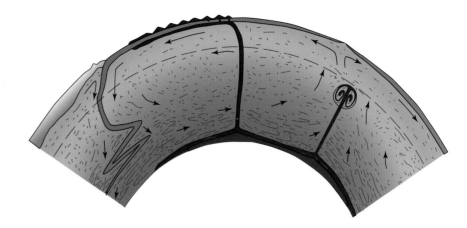

FIGURE 2.2 Sketch of mantle convection and structure based on inferences from fluid mechanics and seismological data. SOURCE: Courtesy of Geoff Davies, Australian National University. After Davies (1999). Copyright 1999 by Elsevier Science and Technology Journals. Reproduced with permission.

Tromp, 1999). However, it is becoming better appreciated that temperature variations may not be the only source of buoyant upwellings in the mantle. Chemical variations may be large enough to affect large-scale mantle flow, and mantle plumes can have both thermal and chemical components to their buoyancy (Davaille, 1999; Farnetani and Samuel, 2003).

Does Convection Occur Through the Whole Mantle or in Layers?

A key question about the modern form of mantle flow is whether convection occurs through the whole mantle

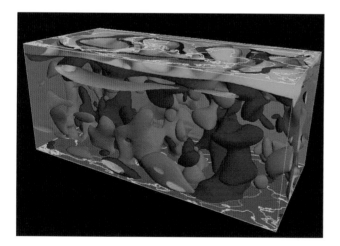

FIGURE 2.3 Representation of large-scale seismic velocity structure of the mantle. Red zones have relatively slow P-wave velocity and blue zones are relatively fast. Slow velocities are thought to represent hotter parts of the mantle. SOURCE: <http://www.seismology.harvard.edu/Projects.html>. See also Su et al. (1994). Used with permission.

or in layers. Models, geochemical analyses of mantle magmas erupted on the surface, and interpretations of seismic waves that have passed through Earth have all yielded different answers. In general, mantle models based on geochemistry suggest that mantle convection occurs in two layers, whereas most geophysical evidence and numerical models strongly support whole-mantle convection. Reconciling these differences is important for understanding Earth's volcanic and thermal evolution.

Geochemical analysis. Interactions of the mantle with the core and surface chemically alter the upper and lower boundary regions of the mantle (discussed below). Convection then stirs this altered material back into the main volume of the mantle. The chemical composition of lavas derived from the mantle provides clues about the extent to which these heterogeneities persist in time and hence about the nature of mantle convection (Van Keken et al., 2002). Lavas (and most other rocks) contain every one of the 90 naturally occurring elements in the periodic table, although about 75 are present in small abundances. With new techniques the concentration of each of the 90 elements and the relative amounts of isotopes of about half of the elements can be measured precisely. The isotopes formed by radioactive decay (^{206}Pb, ^{207}Pb, ^{208}Pb, ^{87}Sr, ^{143}Nd, ^{230}Th, ^{226}Ra, and others) provide detailed information about mantle evolution as well as the processes that produce and transport magma.

Low-abundance trace elements and isotopes of Pb, Sr, Nd, Hf, He, and Os show large, nonrandom variations among volcanic rocks. Basalt lavas erupted at midocean ridges differ systematically from those erupted at hot spots. Midocean ridge lavas also vary from ridge to ridge and along individual ridges (Figure 2.4). The chemical differences between hot spots and midocean ridges have long been considered evidence that the lower mantle (whence mantle plumes presumably come) is different, and convects separately, from the upper mantle (Hofmann, 1997).

Nevertheless, there are complications in the isotopic data. For example, [3]He data suggest that parts of the mantle are relatively unaltered, or at least less degassed, but other isotopes (Nd, Sr, Pb, Hf) tell a different story (Moreira et al., 2001). The mantle overall does not seem to have an Nd or Hf isotopic composition that properly complements that of the continental crust. Many such chemical clues must be sorted out before we can develop a model for mantle convection and the mantle-crust system that agrees with models for Earth's bulk composition derived from meteorites (Question 1) and with the distribution of heterogeneities at depth.

Seismic interpretation. The most direct observations available for inferring the present-day configuration of mantle convection are provided by seismicity in subduction zones and three-dimensional seismic tomography models of the interior. Seismic velocity variations are caused by changes in pressure, temperature, composition, and mineral alignment, so interpretation of the models requires information from mineral physics (Question 6) and geodynamics. High-seismic-velocity features corresponding to cold sinking oceanic lithosphere are clearly observed in regional and global seismic tomography models (Figure 2.5). Low-velocity features (presumably signifying

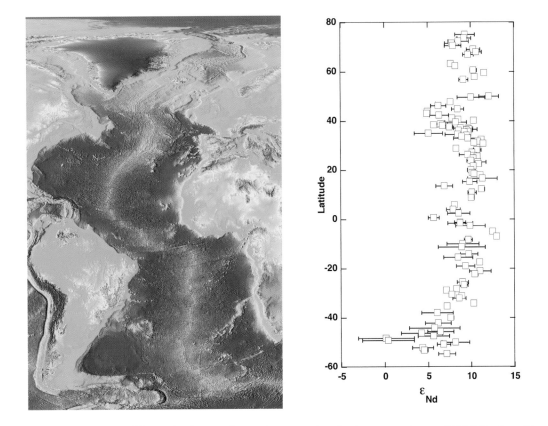

FIGURE 2.4 (Left) Bathymetry of the Mid-Atlantic ridge and topography of adjacent continents. SOURCE: <http://www.ngdc.noaa.gov/mgg/image/2minrelief.html>. (Right) Variations of neodymium isotopic composition in basalt lavas from along the Mid-Atlantic ridge plotted against latitude. Zero on the epsilon scale corresponds to the bulk Earth value, which assumes Earth has the same Sm/Nd ratio as chondritic meteorites. The high degree of heterogeneity indicates that diverse materials are generated in the mantle by melting and subduction and that these heterogeneities are not homogenized by convection. SOURCE: Data from the online database PetDB, averaged by ridge segment by Su (2002).

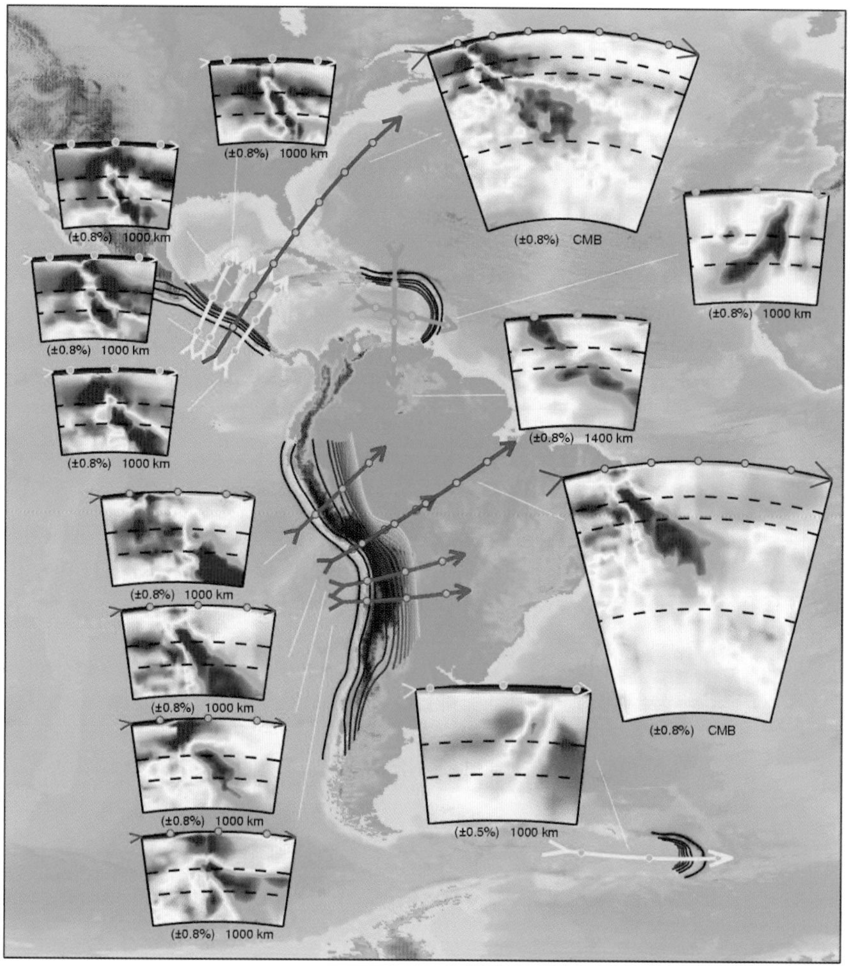

FIGURE 2.5 Seismic tomography data indicating that in some areas subducted slabs extend through the 660-km discontinuity and well down into the lower mantle. Blue shading indicates higher seismic body wave (P) and shear wave (S) velocity, both of which should correlate with lower temperature. The thickness of the cold slab, however, is only about 50 to 100 km, whereas the thickness of the high-velocity (blue) zone is close to 500 km in the lower mantle. The greater thickness in the lower mantle could be due to deformation of the slab or to decreased spatial resolution of the image at greater depth. SOURCE: After Trampert and van der Hilst (2005). Copyright 2005 American Geophysical Union. Reproduced with permission.

relatively high temperature) underlie ocean ridges, back arc basins, and tectonically active areas of continents. Continental cratonic areas are underlain by high-seismic-velocity regions extending 250 to 350 km deep, indicating fundamental differences between oceanic and continental plates (Question 5). Deeper seismic-velocity structures are less easily related to surface tectonics, with very large scale structures tending to dominate in the transition zone from 410 to 660 km deep, and in the lowermost mantle above the core-mantle boundary. For several decades the resolution of seismic tomography models has been improving, and this is guiding numerical modeling of mantle flow processes.

Seismic evidence shows a large velocity discontinuity 660 km below the surface, which is thought to involve mineral phase transformations that tend to impede flow through the transition depth. However, seismological data also show some subducted slabs extending to depths greater than 1,000 km (Figure 2.5),

clearly penetrating the 660-km boundary. The variable depth of lithospheric slab subduction is not easily understood in the context of simple thermal convection and is the primary observation driving consideration of more complex convection models involving both thermal and chemical effects.

Models. Mantle convection models have progressed from simplified two-dimensional models to complex three-dimensional simulations, in concert with increasing computing power and improving knowledge of mantle material properties (Figure 2.6; see also Cohen, 2005). Comparison with seismological models allows some parameters in convection models to be tested, but many issues are still unresolved. Among the challenges of simulating mantle convection are the strong dependence of viscosity on temperature and composition, mineralogical heterogeneity in the mantle on both large and small scales, departures from simple fluid behavior,

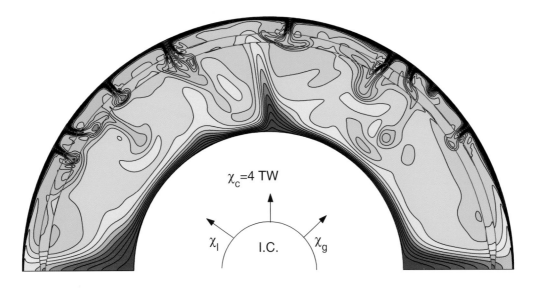

FIGURE 2.6 Computer simulation of mantle convection in two dimensions. Red-green-blue color scale depicts temperatures from 4000°C to 0°C. Fine-scale features, which arise from extreme variations in material properties at small length scales, are not well represented in this simulation, but hot upwellings from the core-mantle boundary region, and cold downwellings (analogous to subduction) from the cold surface boundary layer, are prominent features. SOURCE: Butler et al. (2005). Copyright 2005 by Elsevier Science and Technology Journals. Reproduced with permission.

and the effects of melting and phase changes on material properties. Although the simulations are guided by observations and experimental measurements, these are often indirect or subject to varying interpretations as a result of the difficulty in specifying material properties at conditions of high pressure and temperature. For example, the uppermost mantle is mostly made of three minerals: olivine, orthopyroxene, and clinopyroxene. In the lower mantle these minerals are transformed by pressure into higher density forms, and the size and even the composition of the mineral grains are poorly known. It is the deformation characteristics of these mineral aggregates that determine the nature of mantle convection. Because the grain size and other properties of the deep mantle have yet to be determined, our ideas about convection in the lower mantle involve large extrapolations of the properties we can determine for Earth materials at lower pressure and temperature (see Question 6).

Numerical simulations of mantle convection show that even with phase transitions inhibiting flow and a viscosity increase in the lower mantle, it is plausible that large-scale transport of material between the upper and lower mantle does occur. All in all, current seismological and geodynamic results tend to favor an intermediate model of mantle convection that is neither strictly layered nor simple whole-mantle convection.

When Did Earth's Inner Core Form?

Earth's thermal evolution is reflected in and strongly influenced by the temperature of the liquid outer core. The fact that Earth's outer core is liquid rather than solid is evidence for the hot origin of Earth, and the fact that the core has not completely solidified over Earth's 4.5-billion-year history means that it has been prevented from losing heat too quickly. Laboratory experiments suggest that the top of the core is about 1500°C hotter than the deep mantle (Figure 2.7). Therefore, heat must be flowing from the outer core into the lower mantle, and the core must be cooling. The core must also be close to its solidification temperature because the inner core is solid. As the core cools, it solidifies from the bottom up, so we deduce that the solid inner core is growing and the liquid outer core is shrinking.

The inner core–outer core boundary must have a temperature exactly equal to the melting temperature of

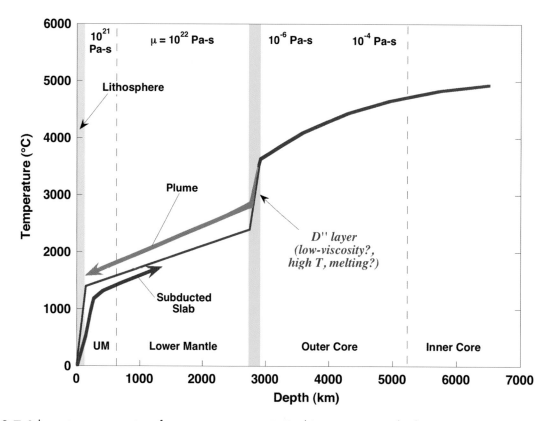

FIGURE 2.7 Schematic representation of average temperature in Earth's interior versus depth. Viscosity estimates are also shown. Temperature is highly uncertain below about 500-km depth. The average mantle temperature (red line) is based on an adiabatic gradient and a temperature of 1350°C at a pressure of 1 bar. Higher and lower temperatures for plumes and subducted slabs are approximate but close to estimates. Temperature in the core and the large temperature drop at the base of the mantle are poorly constrained. See van der Hilst et al. (2007) for a recent estimate of temperature at the core-mantle boundary and Bunge (2005) for a discussion of nonadiabatic temperature structure in the mantle.

the core at the corresponding pressure. The core melting temperature is uncertain because the core contains minor elements other than Fe, and it is not known exactly which elements and how much of them. Hence, the melting temperature of the core is likely to be a complex function of both composition and material properties at high temperature and pressure. Ongoing research is examining the possibility that heat-producing elements (e.g., potassium) may be present in the core and may contribute to a slowing of core cooling.

How long the inner core has existed, its rate of growth, and why the core has not fully solidified are fundamental unresolved issues (Butler et al., 2005). Part of the answer seems to be that the core has been kept in a molten state by the mantle, which because of its much higher viscosity does not remove heat fast enough. Also, once crystallization of the inner core

started, it would slow cooling of the core because crystallization releases heat. It has recently been inferred from convection models that the inner core may be relatively young; it may have begun forming about 1.5 billion to 2 billion years ago (Labrosse et al., 2001). This idea, however, is inconsistent with theoretical models that suggest the presence of a solid inner core may be important for the strength of the magnetic dipole field and for the occurrence of reversals. Moreover, there is evidence that Earth's magnetic field is older than 2 billion years (Tarduno et al., 2007). This apparent conundrum may be partly a consequence of our still poor understanding of the characteristics and processes near the core-mantle boundary, including the values of the temperature contrast and the amount of heat flowing across the boundary (e.g., Bunge, 2005).

How Has Earth's Magnetic Field Evolved Through Time?

It has long been recognized that the main part of the geomagnetic field is sustained by fluid motions in Earth's electrically conducting outer core. These motions cause the magnetic field to change over many timescales, from diurnal to annual to geological timescales. However, a unified picture of how the geodynamo and the core fit into the Earth system has not yet emerged. Important questions about the internal operation of the geodynamo and the relationship between the geodynamo and other Earth processes remain unanswered. These include: How do the geodynamo, mantle convection, and plate tectonics interact? What role did the geodynamo play in Earth's early history? The age of the magnetic field is of interest because the magnetosphere may help keep Earth habitable. For example, the magnetosphere may have been necessary to help Earth retain its atmosphere against the eroding powers of the solar wind, and it partly shields Earth's surface against radiation from space. How important the latter is in preserving life or in modulating the rates of evolution is not agreed upon. New insights on these questions will come from continued satellite and ground-based observations of the geomagnetic field and the paleomagnetic field, dynamical interpretations of the core's seismic structure, and sophisticated numerical dynamos (e.g., Figure 2.8) and models of core evolution.

FIGURE 2.8 A snapshot of a three-dimensional computer simulation of the geodynamo. The magnetic field is illustrated using lines of force; blue lines represent the inward directed field and yellow lines represent the outward directed field. The field is intense and complicated in the model's fluid iron core, where it is generated by fluid motions. Like Earth's field, the simulated field has a dominantly dipolar structure outside the core. SOURCE: Courtesy of Gary Glatzmaier, University of California, Santa Cruz; and Paul Roberts, University of California, Los Angeles.

What Are the Chemical Consequences of Mantle Convection?

The mantle interacts with Earth's surface environment through volcanism, heat and mass exchange at midocean ridges, and subduction. The mantle may also exchange material with Earth's outer core. Overall, the mantle mediates a grand-scale circulation of materials that may extend from the core-mantle boundary to the surface and back again. The nature of this mantle circulation and the processes that produce the interactions at the mantle boundaries are critical to understanding how Earth's chemistry is continually modified. For example, volcanism builds oceanic and continental crust (Question 5) and releases to the atmosphere water, carbon dioxide, and other gases, continually renewing the oceans and atmosphere. Mountain building, erosion, and subduction, which also reflect the effects of mantle convection, remove these same materials and tend to recycle them into the deepest parts of the mantle. At the core-mantle boundary we infer there is mainly heat exchange, but there is tantalizing evidence of chemical interaction as well (Brandon et al., 1999). Still unknown are whether the processes that mediate these exchanges were different in the past. An interesting possibility is that the nature of continents and oceans that support a habitable surface environment today reflect only a particular phase of Earth's cooling and hence might have been absent or much different in the past and might also be much different in the future.

BOX 2.2 Midocean Ridge Hydrothermal Systems

During seafloor production at midocean ridges, magma rises under the ridge to within a few kilometers of the ocean floor. This shallow heat source, combined with the faulting and fracturing that accompany seafloor spreading, causes large volumes of ocean water to circulate down into the oceanic crust, be heated, and then rise rapidly and return to the ocean at the ridge. On its way through the oceanic crust, the water reacts chemically with the rocks, changing their mineralogy and exchanging chemical elements. The hot water that returns to the ocean holds a much different suite of dissolved constituents than the seawater that enters the rocks on the ridge flanks.

About 25 percent of Earth's internal heat that is being lost to the surface today is carried by the water flowing through the oceanic crust at and near ocean ridges. The chemical exchange is large and has a global impact on the chemistry of the oceans. The best-known expressions of the high-temperature part of ridge hydrothermal systems are the "black smokers," which are mounds of sulfide minerals that form where hot (>350°C) fluids enter the ocean. However, most chemical exchange occurs farther away from the ridges, where the fluids are much cooler and more difficult to detect as they reenter the ocean.

The chemical exchange at midocean ridges tends to add CO_2 to the atmosphere-ocean system and make the ocean more acidic (Question 7). These effects are generally balanced over the long term by weathering of continental surface rocks, which tends to remove CO_2 and make the ocean more alkaline, with rivers carrying the alkalinity to the oceans. The balance between hydrothermal acidity and river alkalinity is affected strongly by the mantle convection system of which the ocean ridge system is a part. The chemically altered oceanic crust is returned to the deep mantle in subduction zones, which allows the hydrothermal systems to affect the chemical composition of the entire mantle.

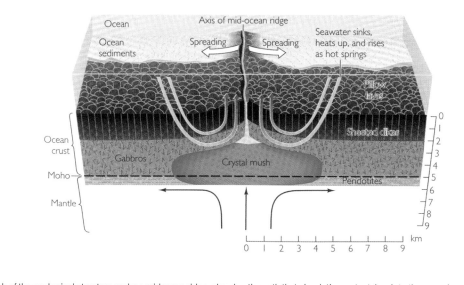

Idealized sketch of the geological structure under a midocean ridge, showing the path that circulating water takes into the oceanic crust on the ridge flanks and back to the ocean near the ridge. The ocean sediment is pealed back in this picture to show the underlying layer of basalt lava. Gabbro is the crystal-line equivalent of basalt, and peridotite is the typical rock of the upper mantle. The partly crystallized magma under the ridge (the "crystal mush"), which has a temperature close to 1000°C, is the heat engine that drives the water circulation. SOURCE: Press and Siever (2001). Used with permission.

Exchange at the surface: Volcanoes. Volcanoes and their associated hydrothermal systems (Box 2.2) provide the primary means by which the mantle passes material to the oceans, atmosphere, and crust. Volcanoes probably created Earth's early atmosphere and oceans, and they continue to resupply these regions with water, CO_2, and other constituents that keep Earth's surface habitable. Volcanoes also produce oceanic and continental crust (Question 5). On some other planets and moons, such as Venus and Io, volcanoes are almost exclusively responsible for the surface morphology. The vast majority of Earth's volcanic activity and crust production takes place at midocean ridges. The current model of ocean

---

BOX 2.3 Volcanic Origin of Oceanic Crust

Earth's mantle melts not because the temperature is raised but because the pressure is lowered as convection carries already hot rock material upward. Melting occurs during slow adiabatic cooling of this upward-moving rock and comes about because the temperature of melting decreases by about 3°C for every kilometer of upward movement, whereas the temperature of the rock decreases by only about 0.3°C/km due to expansion. The temperature and composition of lava erupted at the surface can be used to estimate the depth and temperature at which melting occurs: 150 to 90 km and 1600°C to 1450°C for especially hot mantle plumes like Hawaii (Ribe and Christensen, 1999) and 70 to 10 km and 1400°C to 1250°C under midocean ridges (Asimow and Langmuir, 2003). As a result of cooling during ascent, lava erupts with a lower temperature, typically between 1100°C and 1200°C.

An essential aspect of melting in most planetary interiors is that mantle must be moving upward to melt. On Earth this happens at midocean ridges, mantle plumes, and subduction zones. Other planets—like Mars and Venus—do not have plate tectonics and hence have neither midocean ridges nor subduction zones. Magmatism on these planets may result entirely from mantle plumes (Ernst et al., 2005).

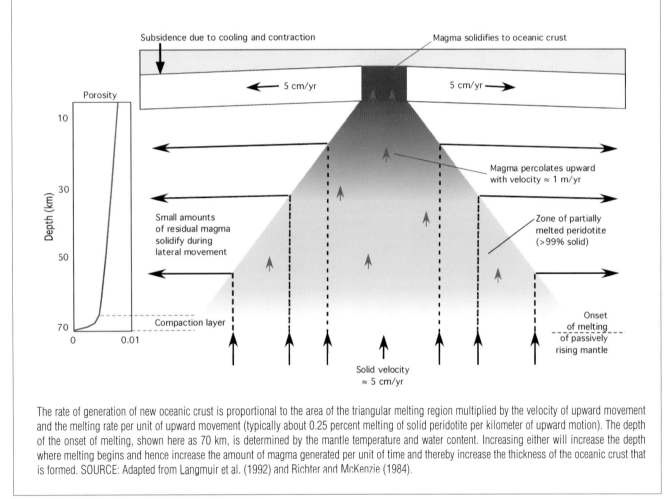

The rate of generation of new oceanic crust is proportional to the area of the triangular melting region multiplied by the velocity of upward movement and the melting rate per unit of upward movement (typically about 0.25 percent melting of solid peridotite per kilometer of upward motion). The depth of the onset of melting, shown here as 70 km, is determined by the mantle temperature and water content. Increasing either will increase the depth where melting begins and hence increase the amount of magma generated per unit of time and thereby increase the thickness of the oceanic crust that is formed. SOURCE: Adapted from Langmuir et al. (1992) and Richter and McKenzie (1984).

---

floor generation by magmatic processes at midocean ridges is a major success of Earth science, explaining the origin and characteristics of 60 percent of Earth's surface and relating both the thickness of the oceanic crust and the depth of ridges below sea level to the temperature of the mantle upwelling under the ridge (Langmuir et al., 1992). The model emerged from a conceptual leap in our understanding of melting in the mantle (see Box 2.3) that, in turn, was made possible by decades of research on the melting behavior of mantle rocks and the percolation of magma through partially molten rock (McKenzie, 1985).

But while magma formation under ridges explains the origin and thickness of the oceanic crust, many other aspects of the process must be resolved before we can fully understand how volcanoes work and extend the ocean ridge model to magma generation in other environments (see Question 9). For example, current models do not explain how magma produced in a broad, 150-km-wide zone under midocean ridges is focused to erupt mainly within a narrow 10-km-wide zone at the ridges. The chemistry and Th-isotope ratios of midocean ridge lavas also do not match model predictions of the depth of the melting region or the way magmas with different compositions and viscosities move and mix under the ridges (Sims et al., 2002). More comprehensive numerical models are beginning to incorporate chemical reactions accompanying magma flow but still suffer from our limited knowledge of the mechanical properties of partially molten rock and our inability to represent the chemical reactions accurately.

A less well understood type of magmatism occurs in association with subduction zones. Although modest in number, subduction zone volcanoes represent nearly all of the explosive volcanoes (Question 9) and the mechanism by which much of the continental crust is produced (Question 5). That volcanoes are located above relatively cold parts of the mantle is evidence that a fundamentally different mechanism(s), perhaps unique to Earth, is responsible for producing magma. Although small-scale convection driven by frictional drag on the slab may cause melting above the slab, water is the melting mechanism invoked most often. Water (in the form of OH^- groups in minerals like amphibole) is carried into the mantle by subducting slabs and then is lost as the slabs are metamorphosed (Tatsumi and Eggins, 1995). This water lowers the melting temperature of the mantle by 200°C or more. If the released water moves upward from the cool slab into hotter mantle above, it can produce the magma needed to generate volcanoes. The supply of water by subduction to the magma-producing regions located 100 km or more below the volcanoes is confirmed by the presence of the short-lived isotope ^{10}Be, derived from the atmosphere, in some island arc lavas. The mechanisms by which water- and CO_2-rich fluids move in the mantle are poorly understood but central to this puzzle; these mechanisms also influence the chemical and isotopic tracers that subducting slabs carry back into the deep mantle. Other processes may also cause melting above the slab, such as small-scale convection driven by frictional drag on the slab.

In general, our knowledge of volcanic processes is much better for near-surface regions than for deeper regions where magma initially forms. A major objective is to understand volcanism from the bottom up—that is, to learn to predict the volume, composition, and eruptive behavior of volcanoes from models of convection and heat transfer processes in the upper mantle and lower lithosphere. The bottom-up approach contrasts with traditional volcanology, which is motivated by hazard assessment to study volcanoes from the top down (Question 9). Bottom-up volcanology may also benefit from studies of other planets, such as Mars and Io, where boundary conditions are different enough from those of Earth to allow models to be tested. Better models for the deep structure of volcanoes and long-term degassing of planetary interiors will require major leaps in our knowledge of partially molten rock and magma, the role of water in melting, the effect of melting on the viscosity of partially molten rock, and the distribution of volatile elements between solids and liquids.

Exchange in the interior: Subduction and mantle plumes. Subduction occurs when old oceanic seafloor moves slowly away from an oceanic ridge and across the ocean bottom, cools, and sinks into the mantle (Question 5). Cold subducted slabs contain rock that has reacted chemically with ocean water (Box 2.2) and sediment derived from continents and shell-forming organisms in the oceans. Although much of the sediment may be scraped off in the shallow part of the subduction zones, the slabs carry some of it, plus chemical and isotopic traces of reaction with the ocean, down into the mantle. In this way subduction changes both mantle geochemistry and the volume and composition of the oceans (Question 7).

The extent to which subducted slabs are assimilated into the mantle is an open question. Some seismological images have high-velocity tabular features in the midmantle and even at the base of the mantle that are suspected to be former oceanic lithosphere. Numerical models indicate that it is plausible that sinking slabs remain cool and coherent all the way to the base of the mantle, where they pile up in a "slab graveyard" (Figure 2.2; Christensen and Hofmann, 1994). If this happens, they could be reheated by heat flow from the core (and their own radioactive elements) and return to the near-surface environment as mantle plumes. There is geochemical support for this notion, which offers a direct mechanism for chemical exchange between the surface and the deep mantle (e.g., Hofmann, 1997; Bizimis et al., 2007). Some models suggest that subducted slabs do not sink that far before being thermally reassimilated by the mantle and that the basal layers of the mantle may be very old and relatively pristine. There is also geochemical evidence for this latter model in the form of high primordial ^{3}He contents of large mantle plumes (Courtillot et al., 2003).

Both models are mute on whether there is chemical exchange between the mantle and core analogous to that between the mantle and the oceans. However, Os isotope data suggest that some mantle plumes contain components that may have come from the core (Brandon et al., 1999). This observation, if confirmed, would be consistent with a deep origin of mantle plumes, although uncertainty remains about whether this core signal could be transmitted through a basal mantle layer, which is both denser and more heterogeneous than the rest of the mantle (Figure 2.9; Garnero, 2000). Whether there is any chemical communication between the core and the lower mantle and what processes could allow this communication to be significant are topics of intense debate (e.g., Scherstén et al., 2004).

Summary

Earth's internal evolution governs much of the planet's evolution as a whole, but because the interior is mostly inaccessible to direct sampling, its study requires a combination of approaches. The seismic waves of earthquakes can be used to determine the elastic properties of Earth's interior, and three-dimensional images of the mantle and core from these waves are being produced at systematically higher resolutions. The structures revealed by seismology are interpreted using new knowledge about Earth materials at high pressure, and great advances have been made in experimental and theoretical mineral physics. We now have sophisticated models for convection in the mantle and core and more precise geochemical and isotopic measurements of mantle rocks. But there are still first-order inconsistencies in the interpretations of available observations, especially for the style of convection and the number and origin of mantle plumes. Recent discoveries of structure and evidence for an unanticipated phase at the base of the mantle have added a new dimension to mantle studies, as knowledge from seismology, fluid dynamics, geochemistry, and cosmochemistry comes together.

Earth's deep interior and surface are connected by volcanism and subduction. Volcanism modifies the internal chemical structure of planets, and great strides have been made in understanding the formation of magma and its transport from the mantle to the surface. But there is still no consensus on many aspects of Earth's magmatic and geochemical history and their relation to surface conditions. For example, we do not know how much of Earth's past volcanism was produced by mantle plumes and how much by plate tectonics, or why there were short periods of intense volcanic activity that could have changed the ocean basins, continents, and even global climate. In addition, we still have only hints about how subduction zones work and how the very existence of plates feeds back on the energetics of the mantle convection system. Finally, we are only now exploring the most fundamental connections between Earth's core, magnetic field, mantle, and surface.

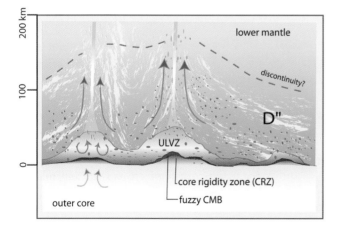

FIGURE 2.9 Inferred features at the core-mantle boundary (CMB). The notation D″ is the seismological designation of the heterogeneous zone at the base of the mantle. ULVZ is ultra-low-velocity zone. SOURCE: Garnero (2000). Reprinted with permission from *Annual Review of Earth and Planetary Sciences.* Copyright 2000 by Annual Reviews.

QUESTION 5: WHY DOES EARTH HAVE PLATE TECTONICS AND CONTINENTS?

Plate tectonics became a central organizing paradigm for geology over 30 years ago. The tenets of plate tectonics theory have been so thoroughly assimilated by the scientific population, and their implications so extensively pursued, that in some ways this report could be considered a description of Earth science in the "post-plate tectonics era." The questions regarding plate tectonics that have now come to the fore have less to do with the soundness of the theory than with the even more basic questions of why Earth has plate tectonics in the first place and how closely it is related to other unique aspects of Earth—the abundant water, the continent-ocean elevation dichotomy, the existence of life. We do not know whether it is possible to have one aspect without the others or how exactly they are interdependent. Can these questions help us understand why Earth is different from the other terrestrial planets?

What Is Plate Tectonics?

Plate tectonics is the representation of Earth's outermost rock layers in terms of a small number of rigid spherical caps or plates. These plates are in relative motion, and their boundaries form the seismic (earthquake-producing) and tectonic (volcanic and mountain-building) belts of the world. The plates interact at three types of boundaries—divergent, convergent, and transform—all marked by the occurrence of earthquakes (Figure 2.10). At divergent boundaries, plates move away from one another as new crust forms between them. The most common type of divergent boundary occurs at the midocean ridge system, which takes the expression of a 40,000-km-long submarine mountain range that rises about 2.5 km above the average ocean floor (Figure 2.11a). At convergent boundaries of oceanic plates, one oceanic plate bends and subducts into the mantle. Convergent boundaries are the loci of the major deep earthquakes (>100 km below the surface); the principal volcanic belts, notably the "ring of fire" around the Pa-

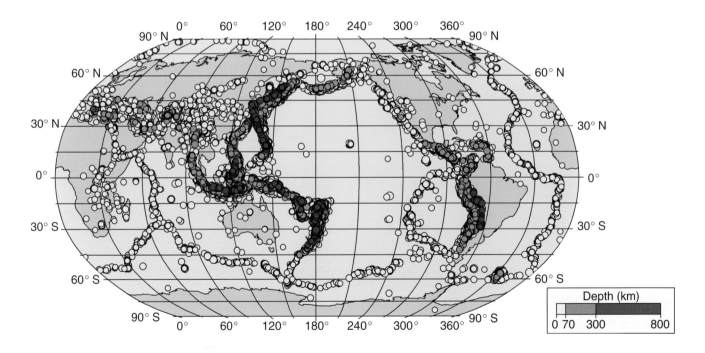

FIGURE 2.10 Locations of earthquakes of Richter magnitude 5 or greater for the period 1991 to 1997. These earthquakes mark the edges of Earth's tectonic plates. SOURCE: Romanowicz (2008). Reprinted by permission from Macmillan Publishers Ltd.: *Nature*, copyright 2008.

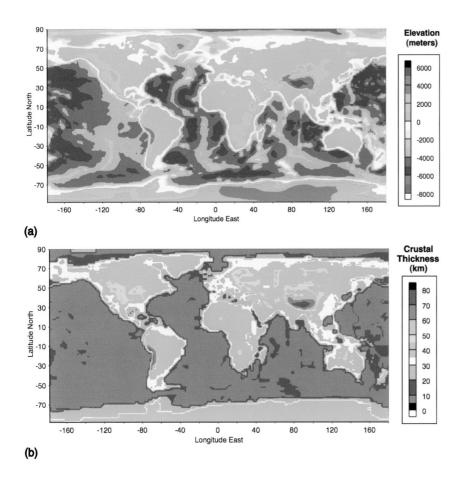

(a)

(b)

FIGURE 2.11 (a) Global topography contoured at 500-m intervals. The preponderance of the continental area lies between 0 and 500 m above sea level. The ocean depth varies with the age of the seafloor. Young seafloor near ocean ridges is only about 2,000 m below sea level. Seafloor that is older than about 60 million years (see Figure 2.12) lies at depths of 5,000 m or more. The elevations shown for Antarctica and Greenland represent the top of the ice sheets; the rocky surface of both areas is below sea level, where the ice is thickest. (b) Contour map of the thickness of the crust. Continents are about 40 ± 5 km thick except in areas of active mountain uplift, where they are thicker, and at their edges, where they are thinner. Oceanic crust is between 5 and 10 km thick, except in areas where there are thick volcanic plateaus. SOURCE: Data from the 2-degree resolution database CRUST 2.0; <http://mahi.ucsd.edu/Gabi/rem.dir/crust/crust2.html>.

cific rim; and mountain building, as in the Himalayas, the Caucasus, and the Alps. At transform boundaries, plates slide past one another, as along the San Andreas fault in California, commonly producing large earthquakes but little volcanic activity.

A key component of the plate tectonics model is the nearly rigid moving plate. The plates are nearly rigid because the rocks near Earth's surface are cool and therefore strong and difficult to deform, even on geological timescales. At greater depths, temperatures rise and the rocks become soft and deformable (Questions 4 and 6). As a consequence, most plates extend to a depth of only about 50 to 200 km below the surface. The relative strength of the plates allows them to move without significant internal deformation. The motion of all points on any rigid plate can be fully described by only two pieces of information: the location of a "pole" about which the plate rotates and the rate of rotation about the pole; this property of rigid plates gives plate tectonics its simplicity and mathematical elegance.

Complexities in the plate model arise from dif-

ferences in the types of crust that comprise the plates. The two types of rocky crust, oceanic and continental, are distinguished by thickness, composition, and age (see also Question 2). Oceanic crust is thin (5 to 9 km), young (less than 200 million years old), and for the most part fairly uniform in chemical composition, consisting of basalt, which is a volcanic rock with silica (SiO_2) content of about 50 percent by weight. In contrast, continental crust is thick (30 to 70 km), varies in age from young to very old (4 billion years), and also varies greatly in composition. The average composition is andesitic, which is a volcanic rock with about 58 percent SiO_2, but locally the composition varies from less than 40 percent to greater than 70 percent SiO_2, with the upper crust being much more silica rich than the lower crust. In general, rock with higher SiO_2 content is less dense, melts at a lower temperature, and is more deformable than rock with lower SiO_2 content. Thirty to forty percent of the radioactive heat-producing elements are concentrated in the continental crust, and as a consequence the deep parts of the continental crust

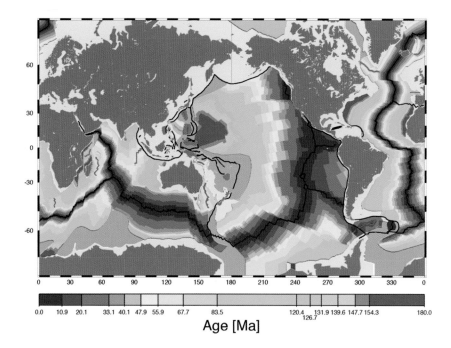

FIGURE 2.12 Map of the age of the ocean floor, with age in million years before present (Ma). Solid black lines are midocean ridges. SOURCE: Müller et al. (1997). Copyright 1997 American Geophysical Union. Reproduced with permission.

are significantly hotter and more deformable than rocks at a comparable depth under the oceans.

The greater thickness, lower density, and more deformable character of continental crust cause it to behave differently than oceanic crust. The different behaviors of oceanic and continental crust influence the nature of plate boundaries. Boundaries that are within oceanic crust tend to be narrow, except in cases where the relative motion between plates is very slow (Royer and Gordon, 1997; Zatman et al., 2001). Boundaries that are between continents tend to be broad because the continental crust is more deformable and much more difficult to subduct, although there is a large variation of deformation styles within the continental crust. Similarly, boundaries that juxtapose oceanic and continental crust exhibit a wide range of deformation styles, from wide to narrow, sometimes changing from one to the other over time. Approximately 10 to 15 percent of Earth's surface is made of broadly deforming regions, while the rest is comprised of the rigid plates characteristic of the plate tectonics model.

The plate tectonics model provides a coherent and simple explanation for many important features of Earth's surface that are not predicted simply by mantle convection. Perhaps its most elegant achievement is to explain the relative youth and other major features

of the oceanic crust, such as why the oceanic crust becomes older, and the ocean deeper, with distance from midocean ridges (Figures 2.11a and 2.12). This age-depth correlation is almost entirely explained by the aging and cooling of the plate as it moves away from the ridge. The oceanic crust is relatively young because it sinks back into the mantle via subduction zones. Plate tectonics also accounts for "continental drift" and allows us to reconstruct where continents were in the past and where they will go in the future. But it still leaves us with significant puzzles about fundamental large-scale features of Earth's crust: the occurrence of hot spots (Question 4), the existence and durability of continental crust, and the complex structure of large mountain ranges where continents collide. It also leaves open the question of why some areas have suffered broadly distributed deformation (e.g., in the Basin and Range Province the distance between what is now Salt Lake City and the West Coast has doubled in about 30 million years), rather than behaving rigidly as is common of plate interiors.

Why Plate Tectonics?

Plate tectonics is a kinematic notion—a description of how things move. Although thermal convection in the mantle (Question 4) has long been recognized as the ultimate driving mechanism (Turcotte and Oxburgh, 1967; Richter, 1973), an outstanding question is why the plate tectonics style of thermal convection extends to the surface and generates the observed plate movement. It was once thought that the plates are pushed apart by convective stresses at the ridge crest. However, the notable feature of the plate tectonics style of convection is that the rocks at the surface take part by plunging back into the mantle at subduction zones. The descent of cold, dense oceanic crust into the mantle at subduction zones is responsible for most plate motions; the subducting rock drags the rest of the plate along with it. Additional small contributions to plate movement come from topography on the base of the plates, but very little comes directly from upwelling of mantle beneath the midocean ridges. The involvement of a buoyant surface boundary layer distinguishes plate tectonics from other forms of planetary convection. Planetary convection would be easier to understand and model if it took place beneath a rigid unbroken surface layer, as seems to be the case for Venus and Mars.

It is easier to understand how plate tectonics can persist once it has started than to explain why it exists, why it exists on Earth but not on other terrestrial planets, how and when it started on Earth, whether it has always been operating or has stopped and restarted at times, and whether it might eventually come to a halt. Plate tectonics would likely be difficult to start, because in order to subduct surface rocks, the cold and strong surface rock layer needs to be broken, the plate needs to bend downward, and once moving downward it must overcome the friction along the boundary with the neighboring plate. All three processes require huge amounts of energy.

The early Earth was much hotter than today's Earth, providing more thermal energy to drive mantle convection. However, the higher temperatures within the mantle would have led to thinner tectonic plates, and perhaps also thicker oceanic crust, making the plates of the early Earth more buoyant than they are today. The net effect on plate tectonics is poorly understood, but it is possible that the Hadean Earth, while almost certainly convecting vigorously, had a style of convection quite different from that of today.

Earth is distinguished from other planets by the presence of significant amounts of water, which may offer a clue about why Earth is the only terrestrial planet with anything that even remotely resembles plate tectonics. It is well documented that trace amounts of water within minerals greatly reduce rock strength and also seem to reduce the ability of faults to resist slippage (Questions 6 and 9). This helps explain how the frictional resistance between plates is overcome in Earth's outer 10 to 30 km. In the deeper Earth, where rocks deform ductilely, relatively low mantle viscosities have been inferred from seismic waves just below the base of the plates (Richards et al., 2001). The low viscosity, or tendency to flow, probably results from water's ability to both weaken minerals and lower the melting temperature of mantle rocks. Computer models can produce a convection pattern that looks more or less like plate tectonics if this low-viscosity zone exists in the upper mantle and if the plate (the near-surface rock layer) behaves as if it were perfectly plastic. But this model still requires that the frictional strength along the boundaries of the plates be several orders of magnitude smaller than that measured on rocks in laboratory experiments, a difference that could be partly explained by the presence of water. Ultimately, uncertainties about the origin of plate tectonics may boil down to better understanding of rock strength on large scales, which is still poorly known for plate-sized rock bodies and for the ultraslow rates at which plates deform.

When Did Plate Tectonics Begin?

It might help to understand the origin of plate tectonics if we could use the geological record to determine when it became the dominant style of convection. This information would be useful because we know that conditions in the early Earth were different from modern conditions (Questions 2 and 4), and hence plate tectonics may have started only after Earth had cooled to a certain degree from its initial state. Most models indicate that the heat flux from Earth was higher by a factor of 3 or more 4 billion years ago and even higher 4.4 billion years ago. In the present plate tectonics regime, the heat flux is approximately proportional to the square root of the rate of generation of new seafloor. If a similar relationship existed before 4 billion years ago, then either the speed of plate motions was at least 10 times faster than today or there must have been a significantly larger number of smaller plates. However, increased heat loss through faster plate creation is potentially problematic in that the plates might not have had sufficient time to cool at the surface until they were dense enough to subduct. The processes responsible for the higher heat loss in the past remain an unresolved issue.

One clue about the beginning of plate tectonics is that granitic rocks, whose formation is presumed to depend on the presence of water at depths of 100 km (see discussion below), had already formed by 4 billion years ago (Question 2). This is indirect evidence that subduction operated in the very early Earth. However, it is also true that granitic rocks can form by means that do not involve subduction. For example, mantle-plume-type volcanism would continually raise new lava to Earth's surface, gradually pushing the older lavas down into the mantle where they would heat up, melt, and produce granitic magma (Richter, 1985).

Other evidence for the early establishment of plate tectonics on Earth comes from geological structures, especially folded and metamorphosed 3.5- to 3.2-billion-year-old rocks. This evidence suggests that lateral compressive stresses and probably large-scale horizontal motions existed during that time, given the old rocks' similarity to those produced along modern plate boundaries. While other forces are also capable of producing horizontal motions and compression without the features of modern plate tectonics, the sum of available data suggests that something like plate tectonics has been the dominant mechanism for shaping Earth's surface since around 3.5 billion years ago and perhaps earlier. The scant rock record from before 3.5 billion years makes it difficult to prove the existence of plate tectonics, but establishing when it started would help us understand the conditions needed for its existence and remains an important objective for both theoretical and field studies of ancient terrains.

What Causes New Plate Boundaries to Form?

Plate boundaries are transient features, so there must be mechanisms to continuously create and destroy them. For example, subduction zones commonly become extinct when two continents collide at a subduction boundary. When microcontinents (or terranes) collide with a continent, subduction can jump from one side of the terrane to the other. Spreading ridges can be destroyed through subduction, as exemplified by the subduction of the eastern extension of the Galapagos rift under South America. Along the eastern margin of the Pacific Ocean, at least one plate (the Kula plate) and most of another (the Farallon plate), along with their plate boundaries, have disappeared beneath North America during the past 100 million years. New subduction zones can be seen to form where transform faults already provide deep cracks in the lithosphere. This appears to be happening today beneath New Zealand, where a transform boundary has been converted to nascent subduction beginning only about 5 million years ago. New spreading ridges apparently form and split continents, as has happened along the Red Sea, where a new region of oceanic crust has been forming between Africa and Arabia over the past 10 million years. The same process appears to be occurring today along the East African Rift, where the northeastern corner of Africa is beginning to move eastward away from the greater African continent. There are many other examples of the formation of new plate boundaries and the termination of old ones. What we know of these examples can be deduced from geological observations, but the causal mechanisms remain elusive.

How Did the Continents Form?

The existence and persistence of continental crust present their own set of questions that are perhaps as fundamental as those of plate tectonics. Continental crust is crucial to Earth as we know it, both because it makes the land surface habitable and because erosion and weathering of the continental surface play a role in regulating Earth's climate (Question 7). But how has the continental crust been preserved at Earth's surface for billions of years, allowing land life to evolve as it has? How were the continents created, and how are they likely to evolve in the future?

Just as water plays a central role in plate tectonics, it also seems essential in "seeding" continent formation. The role of water in continent formation begins at subduction zones, where vigorous volcanic activity tends to occur. These zones produce thick lava accumulations called island arcs that stick up above sea level and are difficult to subduct. The location of subduction zones may well be determined by the presence of sufficient water in rocks to weaken them. It seems inescapable, based on our knowledge of the melting behavior of rocks, that the formation of island arc volcanoes requires water to be carried down within the minerals in the subducting slab. Furthermore, transporting especially large amounts of water into the mantle via subduction zones can help produce the high-SiO_2 magma needed to initiate continent formation. So there is a strong suspicion that water begets subduction, which begets plate tectonics and continents. The planets that do not have plate tectonics—Mercury, Mars, and Venus—have virtually no surface water and probably very little water dissolved in mantle minerals.

If it is true that subduction in the presence of oceans (or water?) inevitably leads to the formation of strips of thick volcanic crust like the modern Aleutian Island chain and the Marianas, and if these strips of crust cannot be subducted, they may be moved around by plates until they collide and combine to build larger land masses, or proto-continents. The bigger these land masses get, the more difficult they would be to subduct. In this way, subduction would act as a kind of filter, allowing thick and buoyant crust to remain at the surface and destroying crust that is thin and dense.

However, this island-arc origin of continents cannot be the whole story because the continents are more silica rich than island arcs. We know that continents undergo geological processing and reprocessing that is too complex to be described in a one-stage model. For example, any land above sea level is subject to erosion and weathering. Weathering tends to leach away "mantle-type" elements like Mg and Ca, as well as others, and generally leaves Si and Al behind. The sediment is transported to the ocean margins, where some is subducted and some is plastered onto the margin of the continent or squeezed between colliding continents. Volcanism also occurs within continents, and the lower part of the crust can itself melt and feed volcanoes in continental margin subduction zones like the Andes and in continental collision zones like the Alps. So continents tend to be repeatedly modified after their initial formation, and they are also broken apart and reassembled by processes related to plate tectonics.

Other processes might also contribute to the particular composition of the continents. In general, if the raw materials for continents have basalt-like SiO_2 concentrations, but the preserved crust has higher SiO_2, then some low-SiO_2 material must be returned to the mantle. This low-SiO_2 material cannot be sediment because sediment is high in SiO_2. There does exist a subduction-like process, called lower crustal foundering, that can return low-SiO_2 crustal rock to the mantle. Seismic studies have begun to document what appear to be large "drips" of high-density material that are slowly falling off the bottoms of continents in some areas (Calvert et al., 2000; Zandt et al., 2004), and studies of xenoliths and lavas also provide evidence that the process has occurred (Ducea, 2002; Gao et al., 2004). The questions that remain are how common lower crustal foundering is, what stimulates it, and how long it has operated. The existence of this continental destruction mechanism is consistent with specific variations in the strength and density of rocks in the lower continental crust and uppermost mantle, but the conditions leading to its initiation must await more detailed knowledge of rock properties (Question 6).

Another poorly understood contributor to continent formation and modification is mantle plumes. In oceanic regions, mantle plumes produce large patches of thick crust with the composition of basalt. But this dense, plume-produced crust could still be sufficiently buoyant to resist subduction and therefore accrete to the continents. Iceland, Hawaii, and the large plateaus of the western Pacific may all be examples of potential new continental crust produced by mantle plumes. Mantle plumes can also deposit their volcanic products directly on (or within) continental crust, thereby adding to the continental mass. And the heat provided by mantle plume magmas entering the crust can cause crustal melting, uplift, and erosion and could even contribute to the breakup of continents. The role of mantle plumes in the evolution of continental crust is a fundamental unresolved issue, one that becomes more urgent and less tractable when considering the oldest continental crust. Whether all continental crust has been produced by island arc volcanism, or whether an alternative mechanism involving wet mantle melting existed early in Earth's history, remains hotly debated but essentially unknown.

How Does the Underlying Mantle Influence Continental Formation?

Continental plates consist not only of crust but also of relatively cold underlying mantle. The peculiarities of this subcontinental mantle provide additional clues about how continents form and persist, but we have not yet been able to fully understand the message. In many places the oldest continental plates reach about 250 km or more in thickness, much greater than the thickest oceanic plates. Evidence suggests that the attached mantle under continents is a melt residue (it has had magma extracted from it in the past) and that the processes that formed it changed fundamentally about 2.5 billion years ago.

The most ancient continental cores, which formed more than 2.5 billion years ago, are areas of prolonged stability. These regions, called cratons, are apparently stronger than surrounding younger regions of the continents that experience periodic disruption. The mantle portion of the cratons is cold and especially thick but has low density due to iron depletion (Jordan, 1988). The iron depletion is difficult to explain unless this mantle was melted extensively, implying melt initiation at high pressures, probably due to the higher temperature of the Archean Earth. The fact that these thick, low-density continental "keels" are found only beneath crust that is older than 2.5 billion years means that they must have stopped forming at that time (Sleep, 2005). Their presence may also account for the longevity of these old patches of continental crust. For example, if they were extensively melted, they would lack water, making them stronger than other parts of the mantle (Pollack, 1986). Alternatively, because earthquakes are confined to the continental crust and do not occur in the continental mantle, the crust may be responsible for the strength of the continental lithosphere. In this scenario the mantle keel limits heat flow into the base of the crust and thus strengthens it (McKenzie et al., 2005). In addition, despite their similar ages, the much more silicic composition of the overlying crust compared to the melts extracted from the keels shows that they are not simple melt-residue pairs. The origin of thick mantle lithosphere under the oldest continental regions and its role in continent preservation remain intriguing fundamental questions.

How Have the Continents Evolved Through Time?

The processes by which oceanic crust is created (at midocean ridges) and destroyed (at subduction zones) are well established. The processes by which continental crust is made are still hazy, and the processes by which continents are destroyed are even less well documented. Yet there is good reason to believe that continental crust is not permanent—simply long lived. The general questions of interest are how the volume of the continents has changed through Earth's history and how the continental volume, ocean volume, thickness of continents, and sea level are related. Corollary questions concern the mechanisms of production and removal of continental crust and whether they have changed with time.

Wholesale subduction of large tracts of continental crust is generally considered unlikely because of its low density and great thickness. However, the thin fringes of continental crust (Figure 2.11b) that surround most continents are not buoyant enough to resist subduction if they are attached to the underlying dense mantle. Calculations, observations in young mountain systems, and the scarcity of deep continental margin rocks in older mountain belts all suggest that subduction of thinned continental crust may be common. Geologists have also discovered rare exposures of continental crustal rocks that were subducted to depths of at least 200 km, recrystallized, and then returned to the surface (Figure 2.13; see Rumble et al., 2003). These so-called ultrahigh-pressure metamorphic rocks bear witness to the subduction of continental crust, but we do not know whether any of the crust remains in the mantle (i.e., that the process effectively recycles continental crust). Continent-derived sediment deposited in oceanic trenches is another source of subducted continental crust. This mechanism of continent removal depends on erosion, and erosion is most effective in areas of high elevation that are produced by continental collisions.

As a result, over Earth's lifetime the total volume of recycled continental crust may be equal to or even exceed the current volume of the continents. Some studies suggest that the volume of continental crust is steady, with the amount of subduction approximately equaling the amount of new crust formed by upwelling magma (von Huene and Scholl, 1991). However,

FIGURE 2.13 Ultrahigh-pressure metamorphic rocks, Dabie Shan, China. SOURCE: Gray Bebout, Lehigh University. Used with permission.

such estimates are imprecise, and there is little evidence that they can be extended to the early Earth. Thus the volume of the continental crust through time, the volume of continental material reworked at subduction boundaries, and the total volume of continental crust subducted remain highly uncertain. Each of these represents a first-order issue for understanding Earth's chemical differentiation. A corollary question regards the extent to which continental material preserved from Archean time (more than 2.5 billion years ago) is typical of Earth's early continents. These remaining bits of ancient continents could have been preserved by chance, but is it also possible that they have unusual properties, such as especially low radioactive heat production (Perry et al., 2006) and the factors discussed above, that made them difficult to destroy?

The fate of continental crust recycled back into the mantle is almost entirely unknown. At depths of about 250 to 300 km, extremely dense minerals like stishovite and hollandite can form, potentially rendering the metamorphosed continental rock more dense than the surrounding mantle, thereby contributing to subduction. Little is known about the phase transitions and metamorphic reactions that might occur at these depths as needed laboratory work has not been done. Hence we have little insight into whether subducted continental crust is returned near the surface, remains in the upper mantle, descends to the core-mantle bound-

ary, or is simply stirred into the mantle by convection. This question has profound implications for chemical cycling and layering within Earth (Question 4).

How Do Climate, Tectonics, and Erosion Shape Landscapes?

A recent advance in geology is the discovery that erosion, precipitation, and mountain building are interlinked in unexpected ways, causing us to rethink one of the most familiar of geological processes. It has long been known that erosion modifies continents, preparing them to be subducted as sediment and enabling mass to be redistributed across Earth's surface. The chemical weathering that accompanies erosion plays a major role in regulating climate (Question 7) and affects the composition of the continents, the oceans, the atmosphere, and the mantle. Erosion affects mainly the rock masses that protrude above sea level and is effective at reducing their elevation down to a value close to sea level. Most of the continental area has an elevation just a few hundred meters above sea level (Figure 2.11a).

The introduction of numerical modeling to mountain-building studies, however, shows that mountain building and climatic processes are coupled. Uplift of mountainous areas is driven by a combination of crustal thickening and erosion and therefore is affected by climate and climate changes. In most mountain ranges, rainfall is higher on one side than the other (Figure 2.14), and hence erosion rates are not strictly correlated with either elevation or average slope. And because mountains influence rainfall patterns, they aid in their own destruction by focusing rainfall onto themselves.

Because the region of maximum erosion in mountain belts may be offset from the region of highest elevation, a complex pattern of mass redistribution can develop (Figure 2.15). In effect, erosion lowers surface elevation but draws the rock upward toward the surface. By this process, precipitation patterns across a mountain range can affect the height, width, and symmetry of mountains, as well as the distribution of fault activity, and can even affect the lateral flow of rock deep in the crust. In other words, deformation and movement of Earth's crust in mountain belts, long thought to be caused entirely by plate tectonic forces, can be heavily influenced by surface processes. This understanding has prompted an intense effort to correlate spatial and

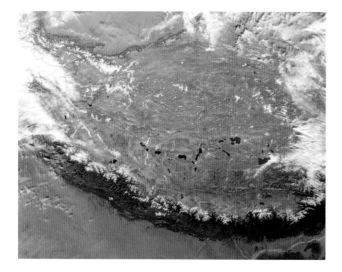

FIGURE 2.14 Satellite photo of the central Himalaya and Tibetan plateau. The strongly asymmetric distribution of rainfall is reflected in the vegetation pattern and distribution of glaciers. The regions with dark green color, on the south side of the mountains, have the highest rainfall and also have the youngest metamorphic rocks exposed at the surface due to the rapid erosion. SOURCE: National Aeronautics and Space Administration, <http://rapidfire.sci.gsfc.nasa.gov/gallery/?2002348-1214/Tibet2.A2002348.0505.1km.jpg>.

temporal patterns of erosion and uplift with varying patterns of precipitation (e.g., Burbank et al., 2003). These studies, in turn, require the latest techniques of measuring erosion rates and crustal movement and imaging the deeper parts of the continental crust and upper mantle. The large scale of tectonic systems has made them challenging to study, a task recently made easier by satellite sensors and systems like Interferometric Synthetic Aperture Radar and the Global Positioning System.

Critical to understanding the coupling of climate and tectonics are empirical models that relate rainfall and topography to the production and transport of sediment and the erosion of bedrock. These geomorphic transport laws are still in their infancy. Experimentally tested, field parameterized, and theoretically sound expressions for most surface processes, especially as they apply to geological temporal and spatial scales, do not yet exist. This gap in erosion process theory presents a great opportunity for scientific advance—and a challenging one because most relevant processes cannot be easily simulated in controlled laboratory settings. Furthermore, the heterogeneity of Earth materials

presents challenges, especially for threshold-dependent processes such as landsliding. New tools, especially cosmogenic radionuclide dating and thermochronology, are now enabling us to determine the rates of processes through space and time, but others will be needed, for example to incorporate biotic effects (Question 8).

The other critical component of models for mountain building, as well as for plate tectonics, is the rheology (deformation behavior) of rocks deep in the continental crust and in the upper mantle beneath the mountains. As Figure 2.15 implies, deep crustal rock flows laterally when pressure is decreased by erosion. The rate of flow depends on the rock properties, which in turn depend on mineralogy, temperature, pressure, stress, and the flow rate itself. Although it is possible to determine the deformation behavior by laboratory measurements, these measurements do not appear to replicate deformation of most rocks under natural conditions. On average, the strength of rocks determined from laboratory measurements is much greater than the strength inferred from the study of regional geological systems (Question 6). This discrepancy is probably a matter of scaling, since natural systems are many orders of magnitude larger, and deform many orders of magnitude more slowly, than laboratory samples. Some large-scale mechanisms of deformation, like faulting, are not reproducible in small-scale experimental samples. Also, fault systems within the crust may self-organize to create high fluid pressure along zones of active deformation, further lowering the stresses needed for continued large-scale deformation (Sleep, 2002). In such cases the strength of deforming rock masses is inversely related to their spatial dimension. Although hypotheses like this one can qualitatively account for field observations, a fundamental theory for the rheology of rocks under planetary conditions and scales awaits development (Questions 4 and 6).

Summary

Although plate tectonics theory explains many of Earth's surface features, fundamental questions remain. There is increasing evidence that the existence of plate tectonics on Earth is related to the presence of abundant water, both at the surface and within Earth's interior, and that water plays a major role in the creation and destruction of continents. However, there is still no

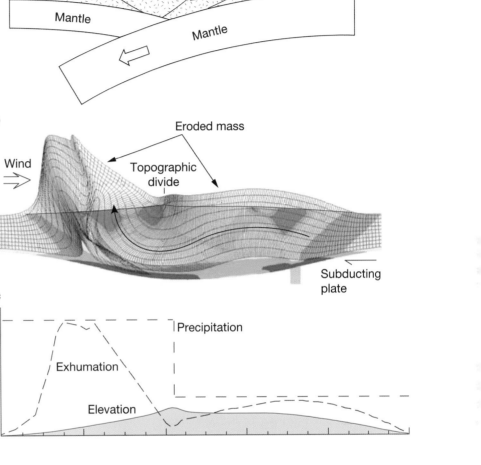

FIGURE 2.15 Links among tectonics, climate, erosion, and topography at convergent plate boundaries illustrated with a hypothetical cross section of convergent plates. The resulting mountain range in (a) is located near the boundary between the plates. A finite-element numerical model (b) assumes stronger rainfall on the left (windward) side of the mountain, which leads to faster erosion there, and general flow of crustal rock toward the region of rapid erosion (curved black arrow). Warmer colors correspond to higher strain rates, the magenta line is the topographic surface, and the gray portion of the mesh shows the eroded mass. (c) Simplified plot of exhumation, elevation, and precipitation for the model. Figure modified from Willett (1999). SOURCE: Dietrich and Perron (2006). Reprinted by permission from Macmillan Publishers Ltd: *Nature*, copyright 2006.

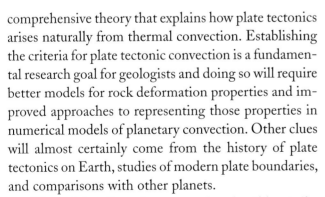

comprehensive theory that explains how plate tectonics arises naturally from thermal convection. Establishing the criteria for plate tectonic convection is a fundamental research goal for geologists and doing so will require better models for rock deformation properties and improved approaches to representing those properties in numerical models of planetary convection. Other clues will almost certainly come from the history of plate tectonics on Earth, studies of modern plate boundaries, and comparisons with other planets.

The origin of continents can be plausibly attributed to the existence of plate tectonics, in particular to the existence of subduction zones. However, the apparent silica-rich composition of the continental crust indicates that the continents are not made in a simple process like that which produces oceanic crust

from magma. Neither the mechanism of producing continental crust nor the process of destroying it and returning it to the mantle is well understood. Nor do we know whether the continents were smaller or larger in Earth's past or whether the processes that produce and shape them were the same. The contribution of mantle plumes to continental formation has gained particular attention, as has the origin of the mantle roots under the oldest parts of the continents.

The past decade has seen a new understanding of the roles of erosion and climate in controlling the structure and shape of mountain ranges. This knowledge has become central to understanding the processes that affect continents and the changes that must be made to plate tectonics paradigms as applied to continental collisions. This search has intensified the desire to quantitatively predict the dependence of erosion rate on other variables and the strength and deformation properties of rocks in the lower continental crust and the upper mantle.

QUESTION 6: HOW ARE EARTH PROCESSES CONTROLLED BY MATERIAL PROPERTIES?

Geology is founded on the central insight that rocks can be read as a record of Earth's history. Rocks and minerals are produced and altered by geological processes—melting, eruption, weathering, erosion, deformation, and metamorphism. Therefore, deciphering the secrets of the rock record begins with an understanding of large-scale geological processes. The keys to understanding these processes are the basic physics and chemistry of the materials that make up the planet. Scientists now recognize that macroscale behaviors—plate tectonics, volcanism, and so on—arise from the microscale composition of Earth materials and indeed from the smallest details of their atomic structures. Understanding materials at this microscale is essential for comprehending Earth's history (NRC, 1987) and making reasonable predictions about how things may change in the future.

The high pressures and temperatures of Earth's interior, the enormous size of Earth and its structures, the long expanse of geological time, and the vast diversity of materials and properties present challenges to investigation. Moreover, minerals are complicated solids that generally contain not only their essential chemical constituents but also trace amounts of almost every element known in nature. Although we can learn much about Earth from the study of pure compounds that approximate real minerals, we also know that even minute amounts of other chemical elements can radically change a mineral's behavior.

Fortunately, the surge of interest in understanding Earth materials at the atomic level has been accompanied by rapid development of new tools, including new synchrotron sources that bring the ability to probe the atomic structure of minerals and liquids (Figure 2.16); high-pressure devices to simulate the distortion of atomic arrangements under huge pressures; and advanced quantum mechanical theory, which promises major advances in our understanding of physics

and chemistry at the extreme conditions of planetary interiors and at the smallest scales of mineral surfaces and nanoparticles. Advances at the other end of the spectrum, when the scale is extremely large and/or the processes are extremely slow, will require advances in experiment, theory, computation, and observation. Only the combination of all four is likely to bring progress.

What Minerals Comprise Planetary Interiors?

As noted in Questions 4 and 5, the nature of the convection and deformation that affect Earth's mantle and crust, and hence models for plate tectonics and Earth's temperature history, depends directly on the material properties of rocks and minerals at the high temperatures and pressures of planetary interiors. The pressure is 136 GPa (1.36 million atmospheres) at the base of the mantle and 364 GPa at Earth's center, while the temperature reaches 4000 K at the base of the mantle and 6000 K at Earth's center (similar to the temperature at the surface of the Sun; Figure 2.17).

Phase transformations. The pressure in Earth's interior is so enormous that it alters the fundamental properties of elements; for example, it can convert insulators to metals and cause magnetism to collapse (Figure 2.18). Such changes occur because pressure compresses and distorts the electron orbitals, thereby changing the most basic properties of the materials. Changing pressures bring about many kinds of phase transformations. The most familiar of these are melting and freezing, but many more complex phase transformations have been identified. Structural phase transitions are also common. The transition from graphite to diamond is well known, but more important for Earth processes is how mantle olivine and pyroxenes change at high pressure.

High-pressure mineral transformations, and their dependence on temperature, allow us to estimate the temperature of the deep Earth and provide constraints on how mantle convection works. Temperatures inside Earth can be estimated by comparing the pressure and temperature conditions at which mineral transformations occur in the laboratory to the depths at which sudden changes in the physical properties of the mantle and core occur (Figure 2.19). We know, for example, that the boundary between the liquid outer core and the

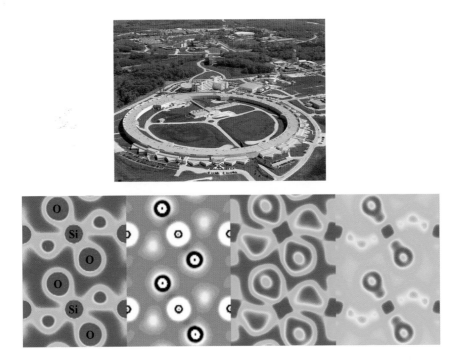

FIGURE 2.16 (Top) Aerial view of the storage ring at the Advanced Photon source. Such third-generation synchrotron sources have revolutionized the study of Earth materials by dramatically increasing spatial and temporal resolution of experimental measurements and allowing for the study of much smaller samples than had been possible. A similar qualitative advance is expected when the first fourth-generation synchrotron sources (X-ray-free electron lasers) come online in 2009. SOURCE: <www.aps.anl.gov/About/APS_ Overview/index.html>. Courtesy of Argonne National Laboratory, managed and operated by the University of Chicago, Argonne, LLC, for the U.S. Department of Energy. (Bottom) Results of a quantum mechanical computation based on density function theory, showing the predicted structure and distribution of electrons in SiO_2 at high pressure. Such computational methods can provide estimates of material properties over the vast range of pressures and temperatures encountered in planetary interiors. SOURCE: Oganov et al. (2005). Reprinted with permission. Copyright 2005 by the American Physical Society.

solid inner core must be at the melting temperature of the core (Question 4), although the temperature is not known precisely due to uncertainty in the composition of the core and the difficulties of exploring these high temperatures in the laboratory. The temperature of the most important changes of seismic wave velocity in the mantle, which happen at depths of about 400 and 660 km, is well constrained by laboratory studies of the conversion of olivine and pyroxene to higher density minerals. These phase transformations are so drastic that they can influence mantle convection; a phase transformation that causes a large change in density can work either for or against the thermal buoyancy that drives convection.

Although the effects of phase transitions on mantle convection are generally appreciated, we still do not know how the natural system actually works—for example, the extent to which the phase transitions impede or enhance the sinking of subducted slabs or change the size and shape of mantle plumes as they rise. A previously unknown phase transformation was recently discovered at pressures well beyond those previously probed (Murakami et al., 2004). The new transformation, from perovskite, the main mineral structure of the deep mantle, to a higher pressure postperovskite form, occurs at the top of the D″ region, an anomalous zone above the core-mantle boundary (corresponding to some 100-GPa pressure) that exhibits intriguing and highly variable seismological features (see Question 4), some of which may be caused by the transformations.

What is the melting temperature of rocks under pressure?
Much of what we know about how Earth's interior works is based on knowledge of the melting temperature of rock and metal, and how this temperature changes with pressure (Question 4). To expand this knowledge,

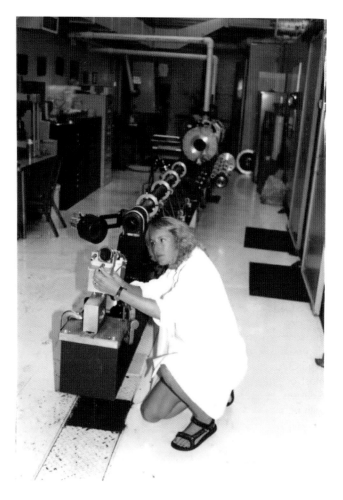

FIGURE 2.17 Diamond-anvil apparatus (top). The sample is placed between two opposed diamond anvils, the tips of which range from 0.01 to 1 mm across, depending on the pressure range of interest. The vertically oriented strip is a metal gasket that prevents the sample from extruding. Diamond is ideal for high-pressure studies because is it strong, chemically inert, and transparent to most light. SOURCE: <http://www.physics. missouristate.edu/Faculty/Mayanovic/mayanovic_research_ webpage.htm>. Used with permission. A shock wave experiment can be carried out using a gun (right), magnetic drive, or laser. The projectile can produce pressures and temperatures that exceed those at Earth's center (like a diamond anvil cell) but for very short periods of time (in contrast to static anvil experiments). New methods combine both static and dynamic approaches to reach pressure-temperature domains (Jeanloz et al., 2007). SOURCE: <www.gps.caltech.edu/~sue/TJA_LindhurstLab Website/index.html>. Used with permission.

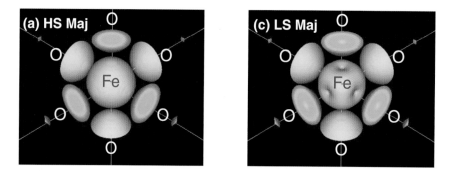

FIGURE 2.18 Influence of pressure on the iron atom. Shown is the predicted charge density of the doubly charged iron cation (Fe) in the mineral ferropericlase (Mg,Fe)O, in which it is surrounded by six oxygens (O). (Left) At low pressure the spins of the d electrons are maximally aligned, producing a net magnetic moment on each iron atom (called the high-spin or HS state) and the magnetic properties that we are familiar with, such as the tendency of magnetic minerals to align with the magnetic north pole. (Right) At high pressures characteristic of Earth's deep mantle the spins pair (called the low-spin or LS state), the atomic magnetic moments vanish, and iron-bearing minerals are nonmagnetic. The figures show that the size and shape of the iron cation also change across the high-spin to low-spin transition: iron is smaller (by about 10 percent in volume) and less spherical in the low-spin state, which should produce a change in density and other physical properties of iron-bearing minerals. SOURCE: Tsuchiya et al. (2006). Reprinted with permission. Copyright 2006 by the American Physical Society.

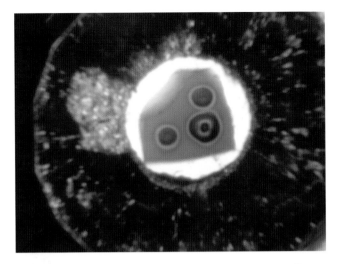

FIGURE 2.19 Photograph looking into a diamond cell at a 100-micron blue single crystal of hydrous ringwoodite (ideally Mg_2SiO_4 composition) held in situ at 30 GPa, corresponding to a depth of 800 km in Earth. The brown spots indicate where the sample has been heated with a laser to a few thousand degrees, causing a phase transformation to the assemblage $MgSiO_3$ perovskite + MgO periclase that is thought to comprise most of Earth's mantle below a depth of 660 km. SOURCE: Courtesy of Steven Jacobsen, Northwestern University. Used with permission.

we need to understand the processes that control the melting and freezing of rocks and minerals in the planetary interior. Melting of rocks involves complex chemistry, because rocks are typically composed of four or more mineral phases, none of which are pure. As rock melts, the composition and density of the liquid portion are different from those of the solid, and thus, with the help of gravity, one can segregate from the other. For example, the lava that erupts from volcanoes is both less dense and compositionally different from the parent mantle rock. Over Earth's long history the repeated processes of melting, melt ascent due to buoyancy, and eruption onto the surface have completely rearranged many of its chemical elements. This process of planetary differentiation, making chemically distinct domains out of a homogeneous starting material, is one of the most fundamental features of planetary evolution (Questions 2, 4, and 5).

One of the more intriguing questions about melting is whether, under some conditions, magma may be denser than the surrounding solid mantle. Magma is highly compressible, so its density must increase rapidly with increasing pressure. The density of solids

also increases with pressure but more slowly. Although there is so far only scant experimental and theoretical evidence, it suggests that magma can be denser than mantle rock deep inside Earth (Figure 2.20; Miller et al., 1991). The consequences of this for Earth's evolution would be profound. If silicate melt sinks instead of rising toward the surface, it could be stored at depth for long periods, where it would be kept hot. The geochemical consequences of this inverted gravitational separation could also be important, but little is known about the distribution of trace elements between solids and liquids at high pressures. Iron-rich liquid would likely exist as a separate, denser phase than Earth's silicate fraction and sink to the center, forming the core (Question 2). The timescale of this descent and the partitioning of elements between the iron-rich and silicate portions during core formation are still uncertain and have profound implications for the chemical composition of the core and the origin of the geomagnetic field (Question 4).

There is confirming evidence that liquid may be present in the deep mantle, especially near the core-mantle boundary. Seismologists have identified thin layers of extremely low shear wave velocity at the base of the mantle, a characteristic of liquid. It has been suggested that this region could be made of dense, partially solidified magma and that it could even be a remnant of the Hadean planetary magma ocean (Williams and Garnero, 1996; see Question 2). If U, Th, and K are concentrated in this deep liquid, it could mean that the base of the mantle produces extra heat from radioactivity, which would affect how we think about the core dynamo and about the overall chemical composition of the mantle. If mantle liquid is in contact with the liquid outer core, it would also mean that chemical exchange across the boundary would be much more effective than if the mantle is solid; this would change the way we think about the origin of chemical heterogeneity in the mantle (Question 4). To resolve these issues we need to know much more about the properties of silicate liquids and solids at very high pressures and temperatures. Recent experimental advances, including measurements of liquid structure in situ at high pressure (Shen et al., 2004), will work hand in hand with theoretical and computer modeling. Modeling of high-pressure properties (Figure 2.20), using the principles of quantum mechanics, shows promise, although at present only a

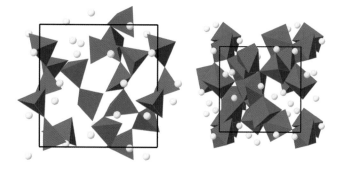

FIGURE 2.20 Predicted atomic-scale structure of a model magma ($MgSiO_3$ composition) showing that the large compressibility of liquids is caused by rearrangement of the structure from an open configuration near zero pressure (left) to a much more compact and highly coordinated structure at the pressure of the core-mantle boundary (right). Silicon-oxygen coordination polyhedra are shown in blue and magnesium ions in yellow. SOURCE: Stixrude and Karki (2005). Reprinted with permission from the American Association for the Advancement of Science (AAAS).

small number of atoms can be modeled, which means that it is not yet possible to use this approach to explore how trace elements behave.

Can seismic waves be used to uniquely determine mantle mineralogy? Material properties and seismology are interdependent in a fundamental way. Seismologists can measure the speed at which seismic waves traverse the mantle and use this information to construct pictures of the deep Earth in a process analogous to a medical CAT scan. At the same time, pictures of the deep mantle cannot be interpreted without information about mantle minerals and rocks, just as radiologists need to know how bone and other types of tissue transmit X-rays. The changes in seismic wave velocity through different structures in the deep Earth are small—about 1 percent—so the elastic properties of the minerals need to be known precisely to interpret the changes. As these properties become better known, geologists hope to use seismic images to map the temperature and composition variations in the mantle and perhaps even the pattern of convection. The latter is possible because seismic wave velocity is dependent on direction, or anisotropy, and can be related to flow patterns if there is sufficient knowledge of the elasticity of minerals and the mechanisms by which they deform (Karato, 1998). A striking example of anisotropy inside Earth may

be seen in the inner core, where longitudinal seismic waves travel 3 percent faster along the rotational axis than in the equatorial plane. This difference may be due to alignment of iron crystals in the core, although the mechanism for producing the alignment is still uncertain (Stixrude and Brown, 1998). Understanding the origin of this alignment is likely to tell us a great deal about the dynamics at Earth's center, the history of the core, and the origin of the geomagnetic field.

How Much Water Is in the Solid Earth?

Earth is unique in the Solar System for its abundant surface water, and most models for the early Earth suggest that the source of this water was the mantle via volcanic eruptions. Based on recent research, it seems likely that the interior continues to be a major reservoir of both water and carbon dioxide (Williams and Hemley, 2001). Earth is so massive that if the mantle is only 0.03 percent water, it would hold the equivalent of all the water in the modern oceans. Upwelling mantle material at midocean ridges appears to contain about this much water, so at present Earth's interior has at least one ocean's worth of water. How much more it might have and how this amount has changed over Earth's history are outstanding questions.

We do not know whether Earth has always had the present amount of water at its surface, but the answer has implications for a variety of processes. To reach the answer, we need a deeper understanding of where water and carbon dioxide are stored in the mantle. We know of two potential reservoirs of water: hydrous phases, such as clays that contain predictable amounts of water within their crystal structures, and nominally anhydrous phases, such as olivine (the most abundant mineral in the upper mantle), which include hydrogen as defects (Figure 2.21). Knowing more about these reservoirs may frame our view of the long-term evolution of the hydrosphere, including formation of the oceans (Question 2). Understanding the evolution of the deep hydrosphere is also central to our view of mantle dynamics, since even small amounts of hydrogen can change the viscosity of the mantle by orders of magnitude and the melting temperature of rocks by hundreds of degrees (Question 4). For example, if the mantle has more water, it might convect faster and produce more volcanism, by which it loses water to the

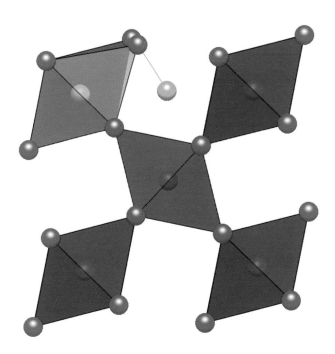

FIGURE 2.21 An example of how water might be stored in Earth's interior. Shown is the predicted structure of a nominally anhydrous mantle mineral (stishovite, ideally SiO_2) with trace amounts of hydrogen incorporated via the replacement: $Si^{4+} = Al^{3+} + H^+$. Dark- and light-blue polyhedra are SiO_6 and AlO_6 coordination environments, respectively; red spheres are oxygen atoms; and the green sphere is a hydrogen atom. The solubility of water in this mineral reaches a few percent at conditions typical of the shallow lower mantle. SOURCE: Courtesy of Lars Stixrude, University of Michigan.

surface. If the mantle loses too much water, volcanism might slow down until enough water is returned to the mantle by subduction. This type of feedback may help regulate Earth's surface environment and the water content of the mantle (see also Question 7).

How Do Minerals and Fluids React?

Chemical reactions between minerals and water enable the oceans and atmosphere to exchange chemicals with the rocks of the crust and mantle. These chemical reactions control the mineral weathering that accompanies erosion and ultimately affect the composition of seawater, the bioavailability of nutrients and toxins in the environment, and the amount of carbon dioxide in the atmosphere (Question 7). All of this chemistry occurs in the microenvironment at the surfaces of minerals. New data about natural materials, especially about the microstructure of mineral surfaces, are changing ideas about how minerals and fluids react. Recent studies have shown that reactivity is exquisitely sensitive to the finest details of surface structure. For example, the rate of exchange with water for oxygen atoms on distinct but structurally similar sites on an aluminum hydroxide surface may vary by seven orders of magnitude (Phillips et al., 2000).

In addition, a major new realization is that most of the mineral surface area in the environment may be in the form of nanophases: extremely small mineral particles, 1 to 100 nm in size, orders of magnitude too small to see with the naked eye. These very small mineral grains have dramatically different physical and chemical properties than larger ones (Banfield and Zhang, 2001). The surface energy of nanophases is so important that it can stabilize structures that do not exist in bulk material (Navrotsky, 2004). These structures may have unique reactive sites, adsorptive properties, and reaction kinetics. The structures of nanophases also vary depending on whether they are surrounded by water, air, or organic ligands. Nanophases are important for their role as a unique reactive surface area, and they also help us understand how minerals form, since all minerals start out as nanophases in the form of small nucleation centers (Figure 2.22).

At and near Earth's surface, the formation and dissolution of minerals take place in the presence of microorganisms, and there is a growing awareness that biology plays a significant role in mediating chemical reactions at mineral surfaces (Question 8). In addition, many minerals are formed entirely by living organisms, both large and small. Limestone, for example, is almost entirely formed as calcium carbonate shell material by small marine organisms. Much of the modern study of mineral formation lies at the interface of biology, chemistry, and geology. With new analytical techniques it is becoming possible to study how minerals are made by organisms and to compare biological and inorganic processes. For example, it is possible that an organism can produce a microenvironment that causes calcite to be precipitated essentially by inorganic processes. By altering the microenvironment, the organism can control the particular form, and hence trace element composition, of the mineral that is precipitated (Bentov and Erez, 2006). We may have much to learn about how minerals form by carefully watching how organisms make them (Figures 2.22 and 2.23).

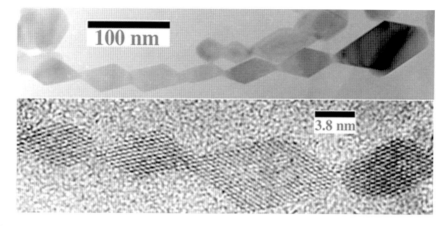

FIGURE 2.22 Necklace of titania nanocrystals that have aggregated spontaneously by oriented attachment. In this mineral growth pathway, crystals no more than a few nanometers in diameter aggregate and rotate so that adjacent surfaces share the same crystallographic orientation. The pair of adjacent interfaces is eliminated and the pair of nanoparticles is converted to a larger single crystal. Individual atoms are visible in the lower view. SOURCE: (Top) Penn and Banfield (1999). Copyright 1999 by Elsevier Science and Technology Journals. Reproduced with permission. (Bottom) Courtesy of Lee Penn, University of Minnesota, and Jillian Banfield, University of California, Berkeley. Used with permission.

Can Large-Domain, Multiscale, and Extremely Slow Earth Processes Be Predicted?

Many properties and processes depend on length scale and timescale in ways that are difficult to predict. The general idea of scaling, or inferring the behavior of materials at one scale from knowledge of those materials at another scale, underlies much of our thinking about Earth. For example, our understanding of mantle convection is founded on our ability to relate planet-scale (large) and laboratory-scale (small) flows that have the same ratio of buoyancy forces to viscous resisting forces (the Rayleigh number). Laboratory analogs are likely to be accurate for some aspects of mantle convection, but they have limits. For example, we know that the crust and uppermost mantle exhibit nonfluid behavior, or there would be no plate tectonics (Question 5). We also know that most of the surface deformation caused by plate tectonics takes place in narrow zones at the edges of the plates. The localization of deformation probably has an origin in complex failure processes that are dependent on both size and timescale. Rocks and even magmas can exhibit a behavior called strain softening, which means that as the amount or rate of deformation increases, the resistance to deformation decreases, which increases the amount and rate of deformation further. Consequently, deformation is most likely to continue wherever it has already started and

to be concentrated in narrow zones rather than being widely distributed. Other feedbacks of this sort include thermal weakening and damage weakening (Bercovici and Karato, 2002). In the latter, deformation either reduces grain size or increases crack density, making the material easier to deform. There are many ways that rocks can behave when stressed; these different deformation processes affect one another; and the larger the rock body under consideration, the more processes that can come into play. Hence, predicting what will happen at a large scale from information about what happens at a small scale is a major challenge.

The behavior of faults raises many scale-related fundamental questions: How are earthquakes (large scale) generated and can we predict them using small-scale models (Question 9)? What localized (small-scale) process and set of conditions trigger a (large) fault to rupture a particular distance on a particular day? How much of continental deformation (large) is caused by slip on faults (small)? Some of the most influential predictors of fault movement have been laboratory measurements of rock strength: squeeze a rock in one direction and eventually it will break or slide along preexisting faults, once friction is overcome. However, rock at the scale of a great earthquake rupture is much weaker than rock in the laboratory. One possible explanation is that water is pervasive in the crust and weakens fault planes by acting as an easily sheared but

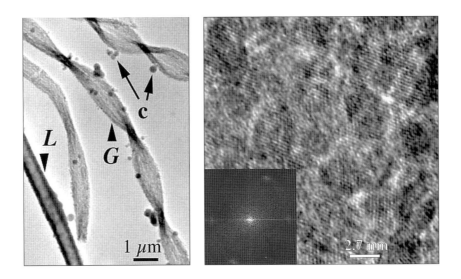

FIGURE 2.23 Orange, polymer-laden ferric iron oxyhydroxides from a submerged mine. The slime consists of colloidal aggregates of nanoparticles, mineralized cell products, and cells (left) of two bacteria. The twisted stalks are characteristic of iron-oxidizing bacteria belonging to the *Gallionella* genus, while sheathed elongate cells are typical of bacteria belonging to the iron-oxidizing *Lepthothrix* genus. The contrast is due to iron oxyhydroxide nanoparticles. (Right) A closeup of the nanoparticle aggregates reveals that while the individual particles are separated (white regions), they have been bio-assembled so that they are crystallographically oriented in the same direction. SOURCE: Banfield et al. (2000). Reprinted with permission from AAAS.

incompressible lubricant that dramatically reduces the friction between the two rock surfaces (Figure 2.24). But as noted above, there are many other possible ways to cause Earth's crust to appear weak in comparison to rocks in the laboratory.

Another reason scaling is challenging is that Earth is heterogeneous: material properties, including viscosity, electrical and thermal conductivity, chemical diffusivity, and elasticity, may vary spatially by orders of magnitude on scales ranging from nanometers to kilometers. Heterogeneity may dramatically influence dynamics. Cappuccino drinkers are familiar with the fluid dynamical oddities of composites, seen in the relative stiffness of milk foam as compared with its constituents, air and milk. Analogous phenomena are common in nature. For example, as magma forms by melting inside Earth, it juxtaposes relatively fluid magma with mineral crystals that are essentially rigid. The viscosity of crystal mush, which largely determines how fast it rises (or sinks), depends strongly and non-linearly on the amount of suspended solid crystals it contains. Deformation and/or dissolution of the solid matrix through which magma moves can also organize solid and liquid fractions so that the liquid becomes channelized, dramatically increasing the rate of liquid-solid segregation, with important implications for

magma migration in the mantle and formation of the core (Questions 2 and 4; Holtzman et al., 2003). The mantle is made of solid minerals with varying strengths. Just as in the case of magma channelization, mantle convection may organize weaker and stronger minerals into layers (foliation), dramatically influencing the viscosity as well as the seismic signal and our interpretation of it in terms of composition, temperature, and flow pattern. Chemical reaction of fluids and melts with surrounding solids can also produce channels, which can significantly influence the composition of the magma and our inferences about its origin (Spiegelman and Kelemen, 2003; Figure 2.25).

The importance of time. The solid-like or fluid-like behavior of the mantle illustrates the importance of time in the material properties of large domains. The boundary between fluid-like and solid-like behavior is set by the Maxwell relaxation time—the ratio of viscosity to shear modulus—which is on the order of 1,000 years for the mantle. This means that we can only determine the viscosity of mantle materials in the laboratory at extremely slow rates of deformation or at unrealistically high temperatures to bring the Maxwell relaxation time within the window achievable by experiment. Just as solids behave like fluids on long

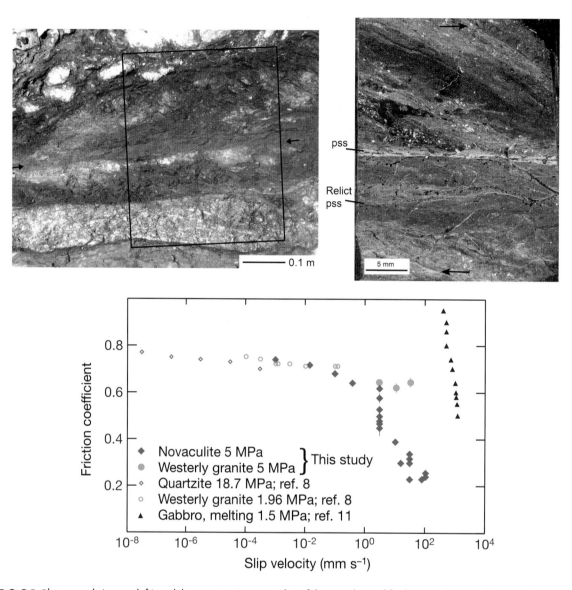

FIGURE 2.24 Photograph (upper left) and thin section (upper right) of the Punch Bowl fault in southern California. The principal slip surface (pss) is thought to have accommodated several kilometers of slip. The slip is localized to a 1-mm (white) region, including a microshear zone with more intense shearing (dark) occurring within a few hundred microns. SOURCE: (Upper left) Chester and Chester (1998). Copyright 1998 Elsevier, reprinted with permission. (Upper right) Courtesy of Judith Chester, Texas A&M University. (Bottom) Results of experiments on fault slip in natural rocks showing that the friction coefficient depends on slip velocity and nearly vanishes for slip velocities similar to those of earthquakes (1 m/s). SOURCE: Di Toro et al. (2004). Reprinted by permission from Macmillan Publishers Ltd.: *Nature*, copyright 2004.

timescales, fluids behave like solids and rupture on short timescales. When magma is deformed very rapidly—for example, during an eruption—it may fracture. Understanding this behavior is helping us sort out the dynamics of volcanic eruptions (Question 9) and how these depend on features such as magma composition (e.g., Gonnermann and Manga, 2003).

Summary

Understanding how Earth works depends on knowledge of the properties of rocks and minerals. After a period of steady progress, breakthroughs are now at hand because of new analytical tools provided by advanced radiation sources (e.g., synchrotron, neutron, and laser facilities) and advanced computing. Much of

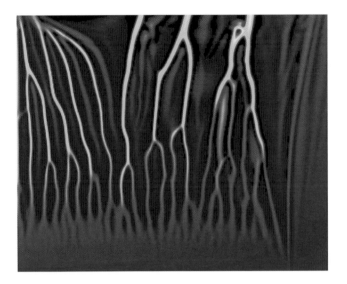

FIGURE 2.25 Simulation of the distribution of melt (as measured by porosity) in a deforming, reacting matrix. The melt organizes itself into channels that vary in width, position, and melt content with time. SOURCE: Courtesy of Marc Spiegelman, Columbia University. Used with permission. See also Spiegelman et al. (2001).

the essential physics and chemistry of Earth materials arises from structures and processes that occur at the atomic level. The new tools allow these small scales to be studied directly as well as simulated, bridging the gap between quantum mechanics and microscopy and paving the way for a new level of understanding of planetary processes at longer length scales.

Earth materials present a challenge to understanding because of their complex chemical composition and the high pressures and temperatures of planetary interiors. The long timescales of geological processes also create difficulties because some of the critical processes that affect planetary evolution take place so slowly that they cannot be simulated in the laboratory and because they may be caused by mechanisms that are not important or even perceptible at laboratory timescales. The physics of large domains, long timescales, and multiple interacting scales remains a major challenge in Earth science and one that will advance only with interdisciplinary effort.

3

A Habitable Planet

Earth's hospitable climate—with temperatures high enough to keep surface water in the liquid state but low enough to keep too much water vapor from entering the atmosphere—is a special and probably critical feature of the planet. There is growing public awareness that climate can change, and there is abundant evidence in the geological record that climate has changed in the past. The history of Earth's climate, a peculiar combination of both variability and stability, poses challenging scientific questions. Our current understanding suggests that many factors can change climate, some capable of producing rapid changes and some requiring much more time but also potentially causing much larger changes. However, despite the many ways that natural forces can change Earth's climate, substantial geological evidence suggests that Earth's overall climate, although it has oscillated between relatively warm and relatively cold states many times, has somehow been maintained in a reasonable, and quite narrow, range that is conducive to the preservation of life. The equitable climate conditions have been present for perhaps 3.5 billion years, despite the fact that both the Sun and Earth have changed in ways that might be expected to play havoc with Earth's climate.

This chapter addresses major questions that relate to understanding how Earth's surface conditions can change and at the same time can be maintained between limits that are conducive to life over extremely long times. Question 7 is concerned with the geological and astronomical factors that affect climate and the geological evidence of climate change. Question 8 considers the relationship between geology, climate, and life. The picture that emerges from Question 7 is that a large number of factors contribute to governing Earth's climate, but how the interplay of these factors results in a particular climate state is still unknown. The answer to this question is critical for addressing future climate change. Question 8 raises the interesting possibility that life itself helps govern climate and other aspects of Earth's surface conditions, while at the same time we have conclusive evidence that climate change has at times been seriously detrimental to life, occasionally killing off huge numbers of species and often forcing evolutionary change.

QUESTION 7: WHAT CAUSES CLIMATE TO CHANGE—AND HOW MUCH CAN IT CHANGE?

Among the systems of planet Earth addressed in this report, climate is the most widely discussed in public forums. We know that human civilizations developed during an unusual period of climate stability over the past 10,000 years or so, but we also know from geological evidence that momentous changes can occur in periods as brief as centuries or even decades in ways that would disrupt human settlement patterns worldwide. Moreover, it is widely recognized that Earth's mean global surface temperature has risen since the beginning of the industrial age and that emissions of CO_2 and other greenhouse gases are at least partly, if not wholly, responsible (IPCC, 2007a). The potentially serious consequences of this global warming, ranging from inundation of densely populated coasts to ocean acidification to the poleward spread of tropical diseases, underscore the need to determine how much of the warming is caused by human activities and what can be done about it. Earth science has an important role in answering both questions.

The immediate grand challenge in climate science is predicting how climate will change over the coming decades. However, the broader challenge is to account for both the long-term consistency of Earth's climate and its multiple and varied excursions in the context of a constantly evolving global geological and biological framework. Only when we are able to capture past climate changes in models will we have confidence in our predictions of future climate. Reliable models have not been available because the conditions that characterized ancient climates—such as ground surface temperature, sea surface temperature, and mean annual precipitation—vanished thousands or millions of years ago, along with the climate they shaped. Lacking real-time data for ancient events, geologists are assembling toolkits of "proxy" data. The temperature and precipitation of continental regions, for example, can often be inferred from evidence left in the sediments of lake beds or in ancient preserved soils. Earth's large-scale surface temperature structure, as well as information on ancient ocean currents, is also reflected in fossil and geochemical records of deep-sea sediments and in records of sea-level change. Similarly, atmospheric temperatures for at least the past 100,000 years or so are recorded in glacial ice and retrievable through deep drill cores in the ice. However, the further we journey into Earth's past, the more different Earth was from our modern planet. To understand Earth's climate in geologically ancient times, we need to know an enormous amount about the geology and geography of the ancient Earth; this is where geological science and climate science become inseparable.

What Processes Govern Climate Change?

The climate system is regulated by how much energy Earth receives from the Sun and how much is radiated back into space (Figure 3.1). How much energy is absorbed depends on the reflectivity (or albedo) of Earth's atmosphere and surface. The albedo depends on how much of Earth's surface is covered by water, land,

or ice; how the continents are arranged; the extent of land vegetation; and the amount of reflective material (clouds and particles) in the atmosphere. It is generally believed that the key determinant of Earth's ability to capture energy from the Sun is the amount of greenhouse gases, predominantly carbon dioxide, present in Earth's atmosphere. Increasing the CO_2 content of the atmosphere stimulates warming, which is then amplified by increasing amounts of water vapor that can evaporate from the oceans at higher temperature. Hence the cornerstone of any broader understanding of Earth's climate is the question of what controls the amount of CO_2 in the atmosphere.

The various processes that contribute to the CO_2 content of the atmosphere are referred to collectively as the carbon cycle. The carbon cycle is a key regulator of climate change. The overarching issue is the fraction of Earth's carbon that is present in the atmosphere in the form of CO_2 or other greenhouse gases like CH_4. For the modern Earth, most of the carbon is stored in rock, and most of that is stored deep within the mantle and core. Estimates suggest there is 500,000 times more carbon stored in Earth's mantle than in the atmosphere (McDonough and Sun, 1995; Salters and Stracke, 2004), and there is likely to be more carbon in Earth's core than in the mantle. Most of the carbon not stored in the mantle and core is found in sedimentary rocks as the mineral calcite or as organic material like kerogen and petroleum. Most of the rest is either dissolved in the oceans, stored in soils, or present as living plant and animal tissue. Only a very tiny fraction (roughly one-millionth) is present in the atmosphere and acting to help warm Earth's surface. The Venusian atmosphere, which contains about 200,000 times more CO_2 than Earth's preindustrial atmosphere, is clear evidence that the distribution of carbon between a planet's interior and atmosphere can be very different from that of Earth.

Even though the amount of carbon in Earth's atmosphere is small, changes in the amount have a major effect on the surface temperature. Although the carbon in Earth's core is not likely to be transferred to the atmosphere, there are ways that at least some fraction of the enormous amounts of carbon in Earth's mantle, crust, and oceans could be. Similarly, there are ways to transfer the carbon in the atmosphere to the oceans and to sediments and then to subduct them into the mantle.

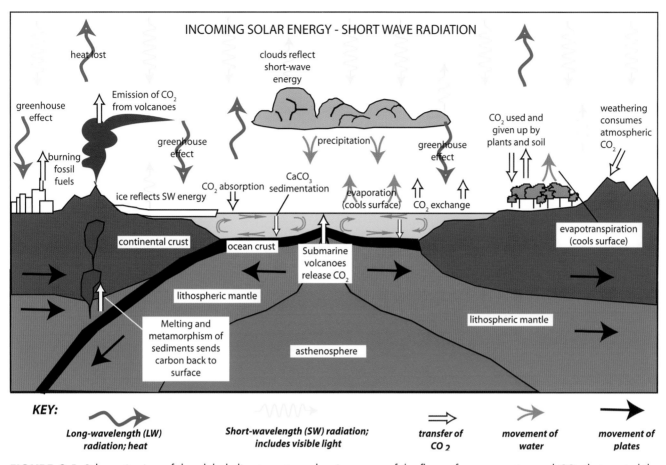

FIGURE 3.1 Schematic view of the global climate system, showing many of the flows of energy, water, and CO_2 that control the climate and the processes that play a role in regulating Earth's greenhouse and determining what happens to the solar energy. Not shown is the warm circulation near midocean ridges, which moves CO_2 from the ocean to the shallow oceanic crust. SOURCE: After <http://www.carleton.edu/departments/geol/DaveSTELLA/climate/climate_modeling_1.htm>. Courtesy of David Bice, Pennsylvania State University. Used with permission.

Studies of the carbon cycle are aimed at understanding how the atmospheric carbon content is regulated by geological and biological processes. Over the past century, fossil fuel burning has overwhelmed natural processes, quickly transferring a large amount of buried carbon (in the form of organic matter, petroleum, coal, and natural gas in sedimentary rock formations) into the atmosphere as CO_2. On longer timescales, natural processes (e.g., volcanism, subduction, chemical weathering, sedimentation, metamorphism, glaciation, wildfires) also shift carbon between the atmosphere, oceans, sedimentary formations, soils, plants, and deep interior. These processes produce cycles of increasing and decreasing atmospheric CO_2 that occur over timescales of thousands, millions, and billions of years.

At each timescale, different processes are primarily responsible for the changes.

Over the past century and through the next, changes in the greenhouse gas content of the atmosphere are the most important factor affecting climate, although changes in atmospheric particulates and clouds are also important. Burning coal, oil, and natural gas continues to add greenhouse gases and aerosols to the atmosphere, reducing emissions of infrared radiation to space and causing Earth's global mean surface temperature to rise. The amount of increase depends on feedbacks in the climate system, especially the (poorly known) feedback from clouds. On even shorter timescales (years to decades), changes in atmospheric particle loading, notably sulfate aerosols, can affect climate, in

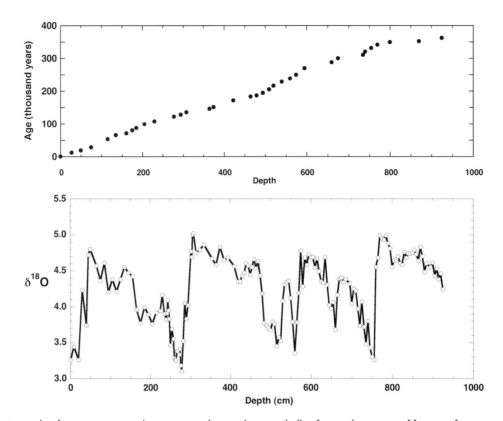

FIGURE 3.2 Example of oxygen isotope data measured on carbonate shells of a single species of foraminifer separated from a 10-m core of deep-sea sediment. Glacial-interglacial cycles are evident. Higher $\delta^{18}O$ values represent times when bottom water temperature was lower and the volume of continental glaciers was larger. Modern time (depth = 0, age = 0) corresponds to an "interglacial" period. Upper graph shows depths where age can be estimated and the estimated age. SOURCE: Data from SPECMAP, <http://www.ngdc.noaa.gov/mgg/geology/specmap.html>.

part countering the effect of increased CO_2. The 1991 eruption of Mount Pinatubo, for example, caused slightly cooler-than-average global temperatures for about a year.[1] Despite the uncertainties and feedbacks, a doubling of CO_2 from fossil fuel burning is now predicted to increase the mean surface temperature 2°C to 4.5°C by about the middle of this century (IPCC, 2007a).

As the period of time under consideration lengthens, more diverse processes that can affect climate come into play. Over thousands of years, variations in Earth's orbit around the Sun (Milankovitch forcing) affect how solar energy is distributed around the globe and lead to changes in mean annual temperature, precipitation, and seasonality. Earth's orbital cycles are responsible in part for the oscillations between ice ages and interglacial pe-

riods that characterize the past 3 million years of Earth history (Figure 3.2). Over thousands of years the oceans are important as well; for example, excess CO_2 in the atmosphere should dissolve into the oceans after about 1,000 years. And in glacial times the increased ice cover on Earth changes the albedo. If ice caps start to grow as a result of cooling over thousands of years, they can reflect more sunlight and enhance cooling.

The role of tectonic processes (volcanoes, mountain building, continental drift) becomes dominant at timescales of a million years or longer (Figure 3.3). Volcanoes, for example, tend to move CO_2 from the deep Earth to the atmosphere, whereas erosion of mountain ranges and the associated chemical weathering of minerals tend to remove CO_2 from the atmosphere and ocean and bury it as calcite and organic matter in sediments on the ocean floor. Plate motions, which rearrange the continents and oceans, affect atmospheric

---

[1] <http://data.giss.nasa.gov/gistemp/2005/>.

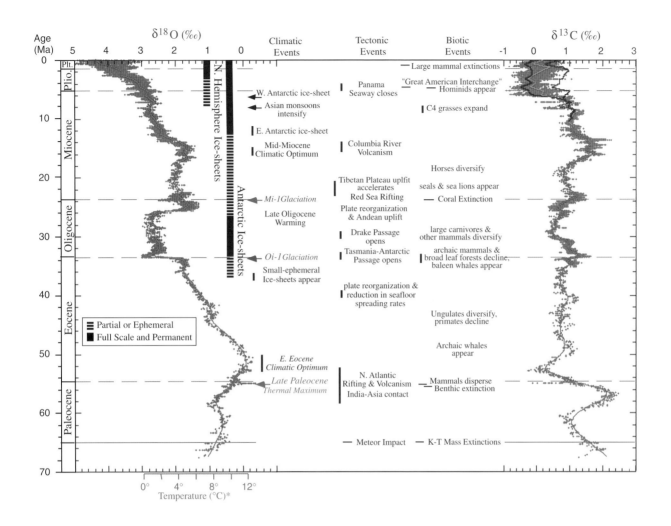

FIGURE 3.3 Global deep-sea oxygen and carbon isotope variations associated with major climatic, tectonic, and biotic events, based on data compiled from more than 40 ocean drilling holes. The temperature scale refers to the temperature of typical water near the ocean bottom and applies only to the time period from 70 million to 34 million years ago. Presently the bottom water temperatures are typically about 2°C and global mean surface temperature is 15°C. Fifty million years ago, bottom water temperatures were about 10°C to 12°C, which corresponds to a global mean surface temperature of about 25°C. From the early Oligocene to the present, about 70 percent of the variability in the δ18O record reflects changes in Antarctica and northern hemisphere ice volume. The vertical bars provide a rough qualitative representation of ice volume in each hemisphere relative to the Last Glacial Maximum, with the dashed bar representing periods of minimal ice coverage (≤ 50 percent), and the full bar representing close to maximum ice coverage (> 50 percent of present). SOURCE: Zachos et al. (2001). Reprinted with permission from the American Association for the Advancement of Science (AAAS).

and oceanic circulation, which in turn changes the efficiency of heat transport from low to high latitudes. These connections can be seen where geological events are correlated with major climate shifts. Volcanic activity that occurred as North America broke away from Europe and the large outpouring of lava that produced the Columbia River plateau about 15 million years ago are both associated in time with globally warm temperatures. The opening of the Tasmanian and Drake

passages as continental drift separated Antarctica from neighboring continents is close in time to the first growth of continental glaciers on Antarctica. Factors on the Antarctic continental shelf, such as the elevation of the Vostok and Gamburtsev Mountain regions, may have played an important role in initiating glaciations as well. Continental drift, combined with volcanism, also closed the Panama Seaway, which once connected the Pacific and Atlantic oceans, drastically changing ocean circulation patterns and probably triggering glaciation in the northern hemisphere about 3 million years ago (Zachos et al., 2001).

Some also think that the continental collision of India with Asia, which formed the Himalayas, the Tibetan plateau, and related mountain ranges, has been a primary cause of Earth's gradual cooling to glacial conditions over the past 50 million years (Raymo and Ruddiman, 1993). The growth of those massive mountain ranges is hypothesized to have accelerated erosion and weathering, yielding dissolved calcium that was carried to the oceans by rivers. This calcium was used by organisms to build shells of calcium carbonate (calcite or aragonite), some of which accumulated as sediment on the ocean floor. This well-known example of sequestering carbon by burying it on the seafloor has drawn broad interest among those searching for a practical way to reduce atmospheric CO_2 (IPCC, 2005). Also, when sedimentary conditions and ocean chemical conditions are right, as has happened many times in the past 500 million years, large amounts of carbon can be held as organic matter within silicate and carbonate sediment on the ocean floor. It is this process that produced the rock formations that we now exploit for fossil fuel.

Why Has Climate Stayed in a Hospitable Range?

The luminosity of the Sun may be an important regulator of climate on timescales of billions of years. Stellar evolution models indicate that the Sun's power output has increased by about 40 percent since it first became a star. The lower solar luminosity 4.5 billion years ago would correspond to an Earth surface temperature about 35°C lower than the present—well below the freezing point of water—if other conditions on the early Earth were similar to those of today (Kasting and Catling, 2003). And yet there is evidence from 3.8-billion-year-old rocks and more controversially from 4.4-billion-year-old zircons (Valley et al., 2002) that the earliest Earth had liquid water at its surface (see Question 2). How can this be possible?

How Earth has remained within the temperature limits for liquid water and life for over 4 billion years is a central question about our planet (Box 3.1). A feedback involving volcanism and weathering may provide a partial answer (Walker et al., 1981; Berner et

---

BOX 3.1 A Hospitable Climate

We know that during the past 4 billion years Earth's climate has varied enough to contribute to the extinction of many species. And yet the variations have been mild enough that life has always rebounded quickly. So how hot is too hot, and how cold is too cold for humans?

Earth's nearest planetary neighbors have both stronger and weaker greenhouse effects, with climates either too hot or too cold for life as we know it. A thick cloak of CO_2 heats the surface of Venus to about 470°C, whereas the thin atmosphere of Mars keeps the mean annual surface temperature at about –56°C (Consolmagno and Schaefer, 1994). In comparison, Earth's mean global surface temperature is currently about 15°C. In our present climate state, the mean annual temperature is about 27°C in the equatorial regions and below freezing (and perennially ice covered) at high latitudes. If we imagine an Earth with a global mean temperature just 10°C higher, the equatorial regions might have temperatures as high as 35°C (depending on how much the tropics widen due to water vapor feedback), unusually hot by human standards; there would be no permanent ice cover in polar regions, and most high-latitude precipitation would fall as rain rather than snow. If the global mean temperature were 10°C lower than today, Earth would be covered with ice to the midlatitudes, more extensively than in the ice ages of the past few million years.

The geological record suggests that climate has stayed within these extremes throughout Earth's history, except for geologically brief "snowball Earth" episodes in the Precambrian. But even much smaller fluctuations in temperature can have a significant impact on human settlement. For example, the Medieval Warm Period (about AD 1000 to 1270) brought extensive drought that may have caused indigenous peoples to abandon the great cliff cities in the western United States (Herweijer et al., 2006); at the same time it made Greenland habitable to the Vikings until the Medieval glaciations of the early and mid-14th century (Barlow et al., 1997). Even with modern technologies, the coldest and warmest areas on Earth support only small populations.

---

al., 1983). According to this model, weathering slows as climate cools, allowing volcanic CO_2 to accumulate in the atmosphere. The added CO_2 warms the climate again, causing weathering to accelerate and prevent further warming. The same feedback loop may have allowed more CO_2 to accumulate in the atmosphere early in Earth's history, compensating for the lower solar luminosity and keeping temperatures above freezing. This stabilizing feedback mechanism would operate slowly and so would be effective only over millions of years; it would not significantly temper the effects of rapid CO_2 increases over the next 100 years. In addition, there is still uncertainty about the effectiveness of this weathering-volcanism feedback because of the competing effect of water–crust interactions as a sink for CO_2 and because of increasing evidence (discussed below) that weathering rates do not depend mainly on Earth's surface temperature. If temperature is not the primary determinant of weathering rates, atmospheric CO_2 could vary rapidly and the fluctuations may be even more difficult to predict because they would depend on global factors such as the rate of mountain building due to continental collisions.

Long-term climate regulation may also involve other processes and other greenhouse gases. During the first half of Earth's history, when atmospheric O_2 levels were low (Holland, 1984; Farquhar et al., 2000), reduced gases may have been more abundant in the atmosphere. Methane (CH_4), for example, could have been present at concentrations of 1,000 ppmv or more, compared to only 1.6 ppmv today (Kharecha et al., 2005). At such high concentration, CH_4 could have contributed 10°C to 20°C of greenhouse warming (Pavlov et al., 2003). Disappearance of much of this CH_4, which must have happened when atmospheric O_2 levels rose at about 2.4 Ga (billion years ago), could explain why Earth became glaciated at that time. This hypothesis is attractive, but it has not been tested directly with data from the geological record. Below we discuss what types of information are available from detailed sampling of this record.

What Caused Exceptionally Warm and Cold Periods in Geological Time?

The geological record of climate change, written in ice cores, sediments, fossils, and rocks, provides clues about how much climate has varied over the past 4 billion years (Box 3.2) and the future habitability of Earth. From this record geologists have been able to identify some of Earth's more extreme climates and the factors that may have triggered them.

One of the warmest extended periods in the geological record occurred in the Cretaceous period, about 120 million to 90 million years ago (Barron and Washington, 1982), when large areas of the continents were flooded with shallow seas (Figure 3.4). At the end

of that time, polar temperatures up to 14°C were high enough to support evergreen vegetation, dinosaurs, turtles, and crocodiles north of the Arctic circle (Tarduno et al., 1998). Equatorial temperatures were 3°C to 5°C warmer than today (Wilson and Norris, 2001), and the deep-ocean temperature may have reached 20°C (Huber et al., 2002) as compared to 0°C to 5°C today. Models and proxy studies suggest that the atmospheric CO_2 concentration during the Cretaceous was 2 to 10 times higher than it is today (Caldeira and Rampino, 1991; Ekart et al., 1999; Haworth et al., 2005), although these estimates are still highly uncertain and we do not know how variable the CO_2 concentration was on shorter timescales during the Cretaceous.

The causes of Cretaceous warming are still unknown. Volcanic activity and hence the input of CO_2 to the atmosphere were probably unusually high, as suggested by the plethora of volcanic mountains and plateaus of that age on the western Pacific Ocean floor. The weathering that removes CO_2 from the atmosphere may have been reduced by two processes: (1) the higher sea level would have reduced the continental area subject to weathering, and (2) this period lacked the major continental collision zones that make mountains, which weather more rapidly than flatter terrain. The paucity of sea ice would also have decreased albedo. The clustering of continents could have changed atmosphere and ocean circulation patterns, increasing the poleward transport of heat and thus making the polar regions warmer relative to the tropics. Whatever the primary causes, the middle Cretaceous is our best example of a greenhouse Earth. However, the geography and ocean circulation are so different today that a future greenhouse may look very different.

The coldest period we know of occurred in the Neoproterozoic. This period is particularly interesting for climate scientists. Conditions then were so drastically different from those today that they strain our understanding of how the climate system works. Between 750 million and 580 million years ago, Earth's surface, including all of the oceans, may have frozen over completely for several brief intervals (Hoffman et al., 1998), creating a "snowball Earth." This hypothesis is vigorously disputed (e.g., Hyde et al., 2000)—not the anomalous cold but its cause, duration, and severity. The cold was almost certainly triggered by transient lowering of greenhouse gas concentrations, and the

BOX 3.2 How Do We Estimate Climate Variables in the Past?

Historical accounts of climate are available for only the past few hundred years, so information about older climate events must be gleaned from alternative archives, such as tree rings and isotopic compositions of ice cores and ocean sediments. These records provide an indirect (or proxy) measure of climate variables, such as temperature and CO_2. Proxies tend to respond to more than one factor in the climate system, so multiple measures are needed to interpret them. The further back in time we go, the fewer kinds of proxy records are available, the more limited their spatial coverage, and the greater the uncertainty in what they mean. Thus, a major effort is being made to expand the collection of proxy observations in space and time and to develop new kinds of proxies (Henderson, 2002).

Proxy Measurements Used to Estimate Climate Variables

Variable	Age Range	Proxy Measurement
Mean temperature	Centuries	Glacier length
Ground surface temperature	Centuries	Borehole temperature measurements
Summer temperature	Few millennia	Tree rings, pollen analysis
Land temperature, precipitation	Millennia	Lake sediments (O isotopes)
Mean annual temperature, precipitation	Millennia	Speleothems (O isotopes)
Sea surface temperature	Millennia	Corals (O isotopes, Sr/Ca, and U/Ca)
Atmospheric temperature	Hundreds of thousands of years	Ice cores (O and H isotopes)
Sea surface temperature	Millions of years	Foraminifera (O isotopes, Mg/Ca)
Land or ocean temperature	Millennia to hundreds of millions of years	Fossils, evidence of ice, sedimentary structures (evidence of water)
CO_2 and ocean pH	Tens of millions of years	Foraminifera (B isotopes, Ca isotopes)
CO_2	Hundreds of millions of years	Soil carbonate (C isotopes), stomatal indices in plant leaves

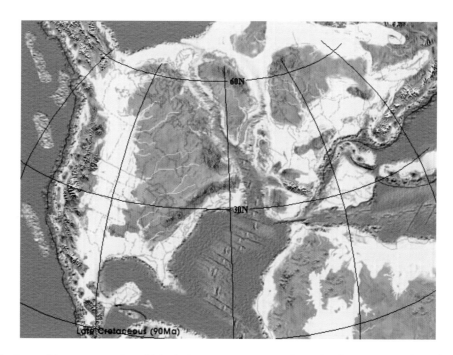

FIGURE 3.4 Physiographic representation of North America, Europe, and North Africa 90 Ma when climate was warm and sea level was high. The land area of the continents was substantially smaller because oceans had risen above the edges of the continents and flooded the interiors. North America was still close to northern Europe, and the North Atlantic Ocean was barely connected to the other oceans. The South Atlantic Ocean (not shown) had not yet formed. SOURCE: <http://jan.ucc.nau.edu/~rcb7/090NAt.jpg>. Courtesy of Ron Blakey, Northern Arizona University. Used with permission.

actual cause may have been the different locations of the continents. The continents were all situated at low to midlatitudes where temperatures are warmest, allowing silicate weathering to proceed rapidly and draw down CO_2 levels, even as the global surface temperature dropped and polar ice accumulated (Marshall et al., 1988; Donnadieu et al., 2004). Alternatively, CH_4 concentrations may have been high during the mid-Proterozoic and then dropped as O_2 levels increased (for a second time; see Question 8) near the end of this time (Pavlov et al., 2003). In either case, as ice cover increased, the albedo and thus cooling would have increased until the planet plunged into an extreme "icehouse" condition. Surface temperatures calculated for this hard snowball Earth are about –20°C at the equator and about –40°C averaged over the globe (Pollard and Kasting, 2004).

The existence of a snowball Earth must be inferred from geological evidence. Translation of such evidence into a hypothesis about Earth's climate and evaluation of the hypothesis using modern climate models and concepts provide an interesting example of the scientific challenges inherent in reconstructing Earth's past conditions. The rock assemblage now considered indicative of the snowball period was initially difficult to decipher. There are marine glacial deposits that formed near the equator, suggesting glaciation in the tropics and hence exceptionally cold conditions; banded iron formations, suggesting anoxic conditions in the oceans; and stratigraphically above and below the glacial deposits there are limestones, which suggest warm conditions (Figure 3.5; Hoffman and Schrag, 2000). In some cases there are nonmarine deposits, which suggest that sea level dropped, and there is carbon isotopic evidence suggesting that photosynthesis all but stopped.

The warm conditions following the snowball Earth period may have arisen because volcanism would have continued through the snowball period, contributing CO_2 to the atmosphere that could not be removed by rock weathering because the rocks were covered with ice. Once extreme levels of CO_2 were reached (~400 times the modern preindustrial level; Caldeira and Kasting, 1992), the greenhouse effect would have been strong enough to overcome the high albedo, melt the ice, and swing Earth to exceptionally warm conditions (~40°C global average in this model) before weathering processes could catch up and remove the atmospheric

CO_2. The temporarily high atmospheric CO_2 would probably have made the rain especially acidic, enhancing chemical weathering and causing a large amount of calcium to be delivered to the oceans by rivers; this may explain the unusual, rapidly deposited limestone layers that cap most Neoproterozoic glacial deposits (Hoffman and Schrag, 2000). A recent three-dimensional climate simulation by Pierrehumbert (2004) has cast doubt on this scenario, however. The new calculations indicate that even 0.2 bars of CO_2 (700 times the preindustrial level) could not have deglaciated a hard snowball Earth. Given the many uncertainties involved in applying climate models to the Proterozoic Earth, it is not yet clear whether the hypotheses or the models are incorrect.

Indeed, there are many arguments against the snowball Earth hypothesis. Even supporters of this theory disagree about significant issues. One is the survival of photosynthetic algae through the plunge in temperatures. How was this possible if the ice was a kilometer thick everywhere as some models have it? Could photosynthetic life have survived in local volcanic hot spots, like modern Iceland? Or did other refuges exist? One variant of the snowball hypothesis, the so-called thin-ice model (McKay, 2000), suggests that the ice in the tropics was only about 1 to 2 m thick, allowing enough penetration of sunlight for photosynthesis. In addition, there would likely be leads and lanes of open water in very thin ice. This model allows Earth to deglaciate at a much lower CO_2 level, only about 30 times the present level (Pollard and Kasting, 2005). However, there are questions as to whether such a solution can be stable, given that sea ice can flow from the poles to the equator, where it would melt (Goodman and Pierrehumbert, 2003). Clearly, much more work is required if the snowball Earth hypothesis is to become an established chapter in Earth's climate history. Nevertheless, even the most moderate of interpretations of the Neoproterozoic evidence for glaciation suggest that it was the coldest period in the past 2 billion years. By comparison, the glaciations that have affected Earth in more recent times have had comparatively little effect on the global carbon cycle.

What Triggers Abrupt Climate Change?

Abrupt climate events are unusual, but they provide insights on the rates at which the climate system is

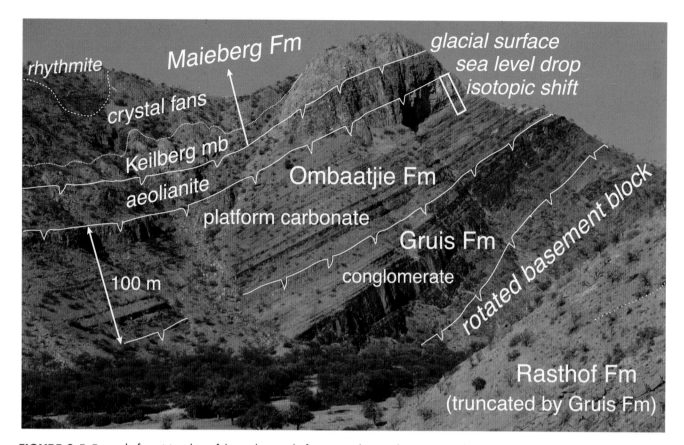

FIGURE 3.5 Example from Namibia of the rock record of extreme climate change in Earth's past. These 750-million-year-old sedimentary rocks have been tilted by tectonic movements, but the time sequence is preserved, progressing from lower right to upper left. The Ombaatjie formation is a limestone deposit formed in shallow ocean water; near its top, isotopes indicate that a glaciation was starting, and above that level the rocks are wind-blown sand dunes, indicating that sea level dropped due to glaciation. Above the dune deposits are limestones deposited after the glaciation ended. The "crystal fans" are a rare type of limestone that is hypothesized to form when inorganic carbonate is rapidly precipitated from the oceans. The time duration represented by this rock sequence is not known, but estimates suggest a few million years. SOURCE: Halverson et al. (2002). Copyright 2002 American Geophysical Union. Reproduced with permission.

capable of change. Much like the extremes of warm and cold discussed above, the rapidity of abrupt climate events provides additional clues about how climate is controlled by Earth processes. Abrupt climate events also serve as important time lines, enabling the correlation and analysis of fragmentary stratigraphic records from around the world. Examples of abrupt climate change include the Permian-Triassic boundary (see Question 8), the Paleocene-Eocene Thermal Maximum, and Dansgaard-Oeschger events in the more recent Pleistocene Epoch.

Dansgaard-Oeschger events, named after the geochemists who first documented them, refer to rapid climate fluctuations that occurred about every 1,500

years during the last ice age, especially the interval between 60,000 and 25,000 years ago (Figure 3.6). Each oscillation is characterized by gradual cooling followed by abrupt warming, typically over just a few decades. Even though these changes are rapid, their magnitude is large—annual temperature swings of up to 16°C are recorded in Greenland ice cores. A number of mechanisms have been invoked to explain them, including solar influences (Bond et al., 2001). Some of the coldest events are thought to be related to massive discharge and melting of icebergs, which would have delivered fresh water to the North Atlantic and possibly changed ocean circulation (reviewed in Hemming, 2004). Similarly dramatic but temporary events almost certainly

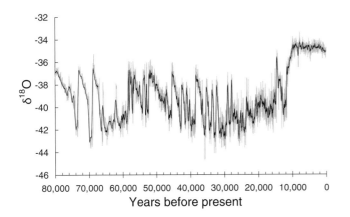

FIGURE 3.6 Record of $\delta^{18}O$, a proxy for mean annual temperature, of Greenland ice from the GISP2 ice core. A change of five units of $\delta^{18}O$ corresponds to a change in temperature of 14°C at the GISP site. The important features of this record are the rapid shifts between 60,000 and 25,000 years ago, when temperatures oscillated by 10°C to 15°C over periods as short as 100 years, and the unusual stability of climate over the past 10,000 years. SOURCE: Data from Grootes and Stuiver (1997).

occurred during earlier glacial periods, although high-resolution ice core and marine sediment records are not available to confirm this.

The most extreme abrupt global warming event recorded in geological history was the Paleocene-Eocene Thermal Maximum, which occurred 55 million years ago (Figure 3.3; reviewed in Zachos et al., 2001). In less than 10,000 years, deep-sea temperatures are estimated to have increased by 5°C to 6°C and sea surface temperatures by as much as 8°C at high latitudes (Stoll, 2006). This warming event was associated with changes in global carbon cycling, oceanic and atmospheric circulation, and the extinction of many marine organisms. Detailed chronology of the interval suggests that it took about 170,000 years to flush the excess ^{12}C from the ocean and atmosphere through burial of carbonate and organic carbon in deep-ocean sediments (Röhl et al., 2000).

The cause of this abrupt event is still debated. At least seven possible triggers have been proposed, including a catastrophic release of 1,050 to 2,100 giga-tons of carbon from seafloor methane hydrate reservoirs (Zachos et al., 2005). A significant shift in osmium isotopes suggests that continental weathering increased substantially (Ravizza et al., 2001), possibly as a result

of increased CO_2 in the atmosphere as well as higher temperature and humidity (Zachos et al., 2001).

Can Earth's Past CO₂ History Be Determined?

The connection between atmospheric CO_2 levels and climate is generally accepted, but there are still few reliable data confirming the relationship through Earth's history. The examples above show that additional or alternative factors, including other greenhouse gases like CH_4, may be required to explain some temperature changes. For example, estimated concentrations of atmospheric CO_2 are too low to explain some of the warmest times of the Cenozoic (Fedorov et al., 2006; Stoll, 2006), and CO_2 concentrations were believed to be very high in the Ordovician and Jurassic, despite evidence of episodically cool climate (Kump et al., 1999; Veizer et al., 2000). Confirming a correlation between periods of warm climate and high atmospheric CO_2 levels during the Phanerozoic remains a major objective. Other key questions include whether other greenhouse gases were important in the more distant geological past, and whether other causes of climate change besides greenhouse gas forcing can be inferred from the geological record.

Much of the work on deep time has focused on proxy studies of marine sedimentary rocks, which record the evolving chemistry of the ocean. Since the ocean and atmosphere are roughly at chemical equilibrium over timescales longer than 10,000 years, and because most of the available carbon is stored in the oceans, reconstructing past changes in ocean chemistry would help establish how atmospheric CO_2 has changed. But because the chemistry of the oceans is so complicated, available data are still insufficient for the task. Further complicating the picture are isotopic data suggesting that steady state models of the carbon cycle are applicable in the Cenozoic (0 to 65 Ma), but not the Neoproterozoic (1,000 to 543 Ma; Rothman et al., 2003). Some aspects of ocean chemistry at least confirm that the oceans undergo major shifts in composition. For example, there is evidence that the ratio of Ca to Mg and the ratio of carbonate (HCO_3^-) to sulfate (SO_4^{2-}) have changed markedly and systematically (Figure 3.7).

Similarly, the rates of past volcanism and weathering cannot be measured directly, and better estimates

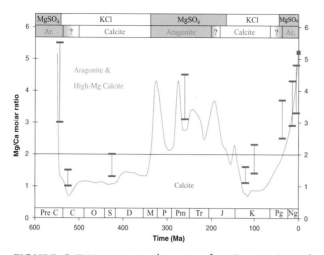

FIGURE 3.7 Variation in the ratio of Mg^{2+} to Ca^{2+} in the ocean over the past 550 million years. Red bars represent values estimated from measurements of fluid inclusions in halite crystals from salt deposits. The gray line is a model. Shown at the top are summaries of geological evidence consistent with the model and measurements. When Mg/Ca > 2, aragonite rather than calcite tends to precipitate from the oceans as the primary nonbiogenic carbonate mineral, and $MgSO_4$, rather than KCl, is the first mineral to precipitate when seawater evaporates to form salt deposits. SOURCE: Loewenstein et al. (2001). Reprinted with permission of AAAS.

are needed to determine how closely they are balanced and how much their changing rates influence the climate system. Volcanism rates are commonly estimated from seafloor generation rates, which themselves must be estimated since most of the ocean floor has been subducted. Seafloor generation rates are calculated from plate tectonics reconstructions and ridge or trench lengths or from global sea level determined from shoreline markers. However, uncertainties are large and results vary. For example, scientists disagree on whether the global rate of seafloor generation has changed over the past 100 million years (Rowley, 2002, versus Engebretson et al., 1992). Interpreting sea-level records is complicated by uncertainties about whether the volume of ocean water has remained constant over the past 500 million years. Figure 3.8 shows deduced sea-level variations for the past 500 million years. The double-humped curve (second column of the chart) has become a backbone of Phanerozoic climate studies and is often regarded as a proxy for the CO_2 supply side of

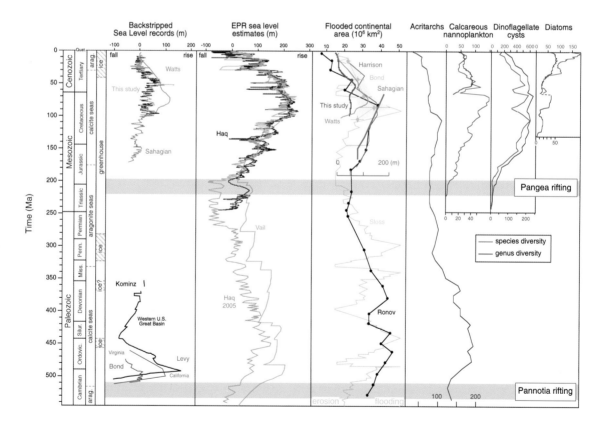

FIGURE 3.8 Amplitude of sea-level change, in meters relative to modern, extracted from the stratigraphic record. The "backstripped" values account for the effects of sediment compaction, loading, and variations in water depth on basin subsidence. SOURCE: Miller et al. (2005). Reprinted with permission from AAAS.

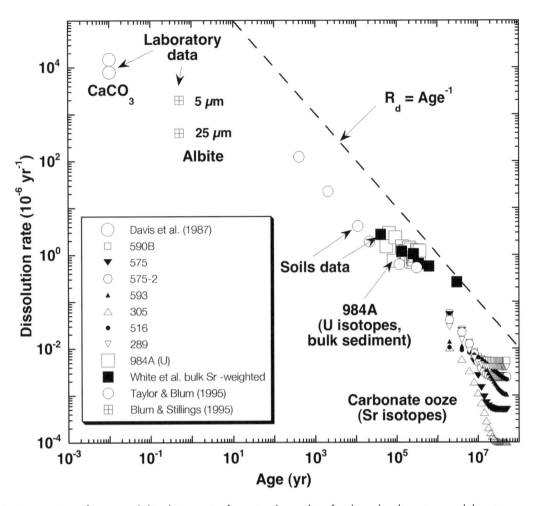

FIGURE 3.9 Comparison of measured dissolution rates for natural samples of soils and sediment versus laboratory measurements. The "age" scale represents the geological age of the material (age of the soil or sediment; length of the laboratory experiment after producing freshly ground powder). SOURCE: Modified from Maher et al. (2004). Copyright 2004 by Elsevier Science and Technology Journals. Used with permission.

the climate equation. Because of this importance, there is a major effort to reduce its uncertainties.

Weathering rates of ancient rock are not well known because of basic uncertainties about the process. For example, newly exposed (fresh) surfaces of mineral grains weather orders of magnitude faster than long-exposed surfaces (Figure 3.9)—a factor that does not appear explicitly in chemical models of reaction kinetics. In other words, areas of active mountain building (e.g., continental collision zones) generate a large amount of fresh mineral surface area by erosion and hence should contribute much more to CO_2 reduction than stable continental areas. Research advances are needed to better understand what controls mineral weathering rates, to quantitatively

relate weathering rates to erosion rates and mountain building, and to evaluate how the age dependence of weathering rates affects models for the regulation of global climate.

A promising proxy for globally averaged rates of weathering in the geological past is the strontium isotopic composition of the oceans. The variation of $^{87}Sr/^{86}Sr$ provides a measure of the relative Sr inflows to the ocean from hydrothermal fluids and eroded continental material, with high $^{87}Sr/^{86}Sr$ indicating a high influx of continental silicate minerals. There is evidence that Sr isotope ratios respond to continental collisions (e.g., Derry and France-Lanord, 1996) and that periods of high $^{87}Sr/^{86}Sr$ are correlated with some glaciations. However, Sr isotope ratios also indicate changes in the types of rocks exposed to constant rates of weathering (e.g., Harris, 1995). The relative contributions of these factors will have to be sorted out before we can determine how to translate the Sr isotopic data into a quantitative estimate of global weathering rates. Other possible proxies for weathering include Os, Ca, and Mg isotopes, although these elements are still in early stages of study.

Summary

The geological record teaches us that Earth's climate has always been changing, but remarkably the surface temperature has remained within a range suitable for life for the past 3.5 billion to 4 billion years. The primary factors responsible for this relatively benign climate are believed to be volcanic emissions of carbon dioxide to the atmosphere, removal of CO_2 by weathering of surface rocks, and more subtle effects, such as the positions of the drifting continents, the patterns of ocean currents, the orientation of Earth's rotational axis and orbit around the Sun, and the luminosity of the Sun. Other chemical and biological effects are also likely to be important, such as the oxidation state of the atmosphere and the concentrations of other greenhouse gases. Interspersed in this vast and mostly life-supporting history are a few periods when Earth was considerably warmer than it is at present, and completely ice free, and a few times when Earth might have been extremely cold and completely ice covered.

At present the greenhouse gas content of the atmosphere is increasing rapidly. The greenhouse gas content of the atmosphere is the most important determinant of climate on geologically short timescales, and models can be used to predict how climate will change over the next decades and centuries. Over longer geological time periods, natural geological processes control the greenhouse gas content of the atmosphere, and other geological and astronomical factors are influential. We have a good qualitative understanding of the factors that contribute to Earth's natural climate states, but we still lack a comprehensive model that can account for the climate changes of the past or predict climate changes into the distant future. Better models for both the volcanic and weathering components of the climate cycle, more quantitative descriptions of erosion and its relation to weathering, and the incorporation of inputs from the biosphere and other factors will likely lead to a more accurate understanding of Earth's climate and climate history.

QUESTION 8: HOW HAS LIFE SHAPED EARTH—AND HOW HAS EARTH SHAPED LIFE?

It is not surprising that many Earth scientists have viewed the geological evolution of Earth as a fundamentally inorganic process—dominated by titanic mechanisms such as mantle convection and plate tectonics. After all, virtually all of Earth's organic mass exists as a veneer of frail and short-lived creatures within a few vertical miles of the outermost surface, a seemingly insignificant afterthought to this massive planetary body of rock. And yet this multitude of organisms—most of them microscopic packages composed primarily of carbon, hydrogen, nitrogen, and oxygen—determines major features of the atmosphere, oceans, and continents. Biologically influenced processes like erosion and weathering, for example, continually shape and reshape Earth's surface. And as we have seen in Questions 4 and 5, the erosion and weathering influenced by life forms affect not only the topography and composition of continents but also the chemical composition of subducted crust and therefore the mechanism of plate tectonics and the composition of the mantle.

Life scientists, in the same spirit, have regarded the evolution of life as a fundamentally biological issue, dependent primarily on time, chance, and competition to trend toward increasing diversity and complexity. We now know that Earth itself is not the mere substrate or background for life's activities as once supposed but rather an active partner in evolution. Geological processes and astronomical events have strongly and repeatedly influenced the story of life on Earth and often determine the kinds of life that can survive and flourish.

The interconnectedness of life and the environment has been a subject of continuing research and debate. An extreme view is that life controls Earth's surface environment and does so in ways that are most beneficial to the continuation of life (Lovelock, 1979). But evidence in the geological record, especially of mass extinctions, suggests that life cannot always maintain conditions favorable to life. We are far from understanding how much of evolution is purely biological and how much has been forced by Earth processes; nor do we know exactly how much of Earth's environment is determined by the presence of life. And yet these questions have suddenly become more urgent as we find ourselves in an era when—presumably for the first time—Earth's surface environment can be manipulated by a single dominant life form, *Homo sapiens*, that is capable of making choices about the effects of its actions.

How Does Life Affect Geological Processes?

Life affects Earth's planetary processes in several ways. At the microscopic scale, life is an invisible but powerful chemical force. Organisms can catalyze reactions that would not happen in their absence, and they can accelerate or slow other reactions. The chemical reactions they enhance have a specific character; in general they extract energy from Earth and from sunlight to fuel life processes. These reactions, compounded over immense stretches of time by a large biomass, can generate changes of global consequence. An example of this global influence is the processing of carbon and oxygen. Weathering reactions on land, combined with organic precipitation of carbonate shells in the oceans, remove carbon from the atmosphere and convert it to carbonate minerals on the seafloor (Question 7). Photosynthesis also extracts carbon from the atmosphere, converting carbon dioxide into oxygen plus organic material. Some of this organic carbon is stored in soils, ocean sediments, and the living biomass of the continents and oceans, while the oxygen is delivered to the atmosphere. Larger animals and plants also have physical effects on Earth, such as promoting soil formation and moderating erosion.

Beyond these generalities, we understand little about the details of biologically mediated chemical processes in the environment, especially those of the distant past. Like many fields of science, however, this one is being revolutionized by powerful new analytical tools and computational techniques. For example, new ultrahigh-resolution microscopes can now be used to observe microorganisms in the environment and in laboratory experiments (Figure 3.10). Synchrotron X-ray techniques can be used to study the chemical processes of these microorganisms. Innovative isotopic techniques are being used to help understand the complicated chemical processing that organisms can achieve. DNA sequencing methods have brought a new dimension to studies of microbiological processes. In the past it was difficult to identify the organisms in natural samples because many could not be cultured. Today, organisms do not need to be cultured; their identity can be determined directly from their DNA. Computational chemistry (see Question 6) also shines a strong new light on natural biochemical processes, bringing the possibility of calculating from quantum mechanical theory how atoms and molecules will behave in the microenvironments surrounding tiny organisms.

Soils represent a particularly clear example of how multiple fields, including inorganic chemistry, physics, and hydrology, can wrest new insights from geobiological processes. Inorganic weathering of minerals and organic carbon in the soil environment releases nutrients and carbon. The rate of release and the types of nutrients define the environment in which life can exist and control the range and abundance of life forms that can survive. In addition, the roots of land plants, as well as bacteria, fungi, and animals such as earthworms, can accelerate the weathering of mineral and organic matter in soils. Such biological catalysis of weathering processes can enhance the suitability of soil for life and

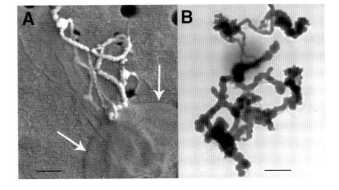

FIGURE 3.10 High-resolution images of (A) a cell (outer cell wall indicated by white arrows) and associated mineralized filaments (white) and nonmineralized fibrals (gray) and (B) FeOOH-mineralized filaments filtered from water. SOURCE: Chan et al. (2004). Reprinted with permission from AAAS.

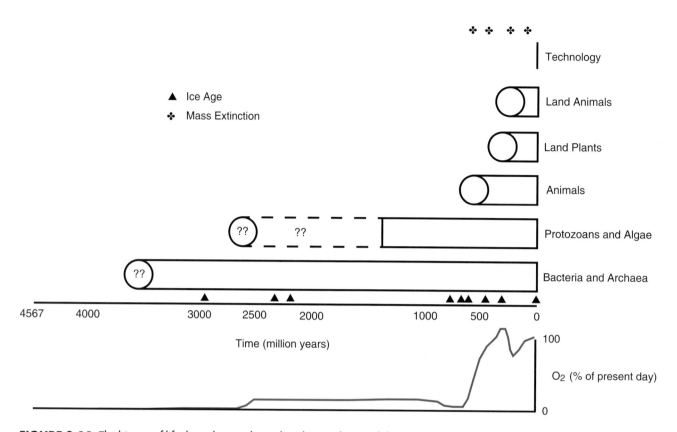

FIGURE 3.11 The history of life, based on geological evidence, along with long-term oxygen, ice ages, and mass extinctions. Molecular data suggest that eukaryotic organisms (protozoans, algae, fungi, plants, and animals) share a common ancestor with Archaea.

also speed up the weathering that would have gone on in the absence of life. The ultimate control on the soil environment is probably climate; insufficient rainfall, for example, limits how fast both inorganic and organic chemistry can proceed. But on a global basis we now know that soil chemistry is powerful enough to affect climate by helping to regulate atmospheric carbon dioxide.

Similarly, we know that vascular plants have an enormous effect on Earth's environment. Life on Earth originated nearly 4 billion years ago, but land plants are found in the geological record only during the past 400 million years or so (Figure 3.11). Several lines of geological evidence suggest that diversifying land vegetation changed the nature of continental weathering, erosion, and sedimentation, changing the physical stability of stream banks and even influencing the composition of Earth's atmosphere (Berner and Kothavala, 2001). Roots break up rock and help transform it into soil. Deep roots also contribute carbon dioxide to soils, resulting in concentrations of soil

CO_2 that are 10 to 100 times higher than the modern atmosphere. The high CO_2 concentrations in soil gas act to acidify soil water, which leads to increased rates of dissolution of minerals. Deeply rooted plants can also extract water from well below the surface and return it to the atmosphere via evaporation from leaves. This evapotranspiration has an important cooling effect on the land surface, as does the shade provided by the leaf canopy.

There is ample evidence that plants and animals also influence erosion rates, but there is still uncertainty about how important they are in the long-term evolution of continental surfaces and how their effects should be represented in new landscape evolution models (Dietrich and Perron, 2006). Erosion itself affects habitat conditions and can strongly influence biodiversity and ecosystem processes. Hence a central question is the extent to which life and landscape evolution are related. For example, do hillslope shapes and river forms reflect the presence of life, or would Earth's land surface be more or less the same shape if the planet were lifeless? A related question is whether topography, which is created by mountain uplift and erosion, affects the structure of ecological communities. The prospect of changing climate in the near future brings up other, possibly more urgent, questions. For example, we would like to know whether rates of erosion will change with changing climate and whether climate-induced variations in vegetation will reduce or enhance the response of erosion rates to climate change. To answer these questions, we need much better models of the effects of biota on weathering, erosion, and sediment transport rates. Biotic diversity needs to be linked directly to changes in material strength (resistance to erosion), mass loss, and sediment mobilization from hill slopes. Similarly, ecological theory needs to include explicit physical effects that influence food web processes. These issues lie at the interface of ecological and Earth sciences. Since the locus of life is typically found in the soil that cloaks the landscape, there is a great opportunity to integrate the fields of pedology, hydrology, geobiology, geochemistry, and geomorphology into a new understanding of this life-supporting system (NRC, 2001).

How Long Has Life Fostered a Habitable Surface Environment?

Because organisms on Earth help maintain a life-supporting surface environment today, it is natural to ask whether they have always done so. This question proves surprisingly difficult to answer, however, largely because we have so little evidence from the early geological record. The parts of early organisms richest in information are organic, namely proteins and nucleic acid, but these are also the most reactive and appealing to a gauntlet of other organisms bent on using them as food. Even biomolecules that reach the seafloor after death are usually broken down by decay processes within the sediments. Therefore most of our paleobiological information is gleaned from the bones, skeletons, and other hard parts preserved as fossils in sedimentary rocks.

For the interval of Earth history that begins with the Cambrian Period (542 Ma), paleobiologists have abundant fossils of plants, animals, and selected algal and protozoan groups that preserve a compelling record of ancient diversity, ecology, and evolutionary pattern. Fossils of microorganisms, including the geologically

important cyanobacteria, provide at least an impressionistic view of evolution and diversity that extends much deeper into our planet's early history (see Question 3). Not all organisms, however, produce the mineral skeletons and tough organic materials preserved as conventional fossils, and this is particularly true of the microorganisms whose metabolic capabilities define much of the interface between the physical and biological Earth.

Fortunately, a new set of tools has become available to establish the presence and infer the biological activities of microorganisms in Earth's history. Most organic compounds in microbial (and, indeed, all) cells decay quickly after death. The exception is lipid molecules found in cell membranes. These hardy compounds, commonly called biomarker molecules, can survive long-term burial in sedimentary rocks and so record aspects of the diversity, environmental setting, and metabolic workings of microorganisms spanning more than 2.5 billion years of our planet's history. Biomarker molecules led geologists to the understanding that petroleum has a biological origin (Triebs, 1936); they have shown how microbial communities responded to transient oxygen depletion in Mesozoic ocean basins (Kuypers et al., 2002), illuminated the nature of life and environments in Proterozoic oceans (Brocks et al., 2005), and provide our earliest evidence for the presence of life's great evolutionary branches in late Archean ecosystems (Brocks et al., 2003). Preserved organic molecules have even been reported as biomarkers in 3.5-billion-year-old rocks from Australia (Marshall et al., 2007). Much remains to be learned about the sources, function, and biosynthesis of biomarker molecules, but new research that combines microbiology, genetics, and emerging technologies for analysis (e.g., Brocks and Pearson, 2005) promises unprecedented insights into evolution and environmental history both in the marine realm (e.g., Grice et al., 2005) and on land (Freeman and Colarusso, 2001).

How Did Organisms Influence the Oxygenation of the Atmosphere and Oceans?

Perhaps the most obvious and vital link between life and Earth systems, at least from a human point of view, is the maintenance of abundant atmospheric oxygen, a feature whose development we still do not

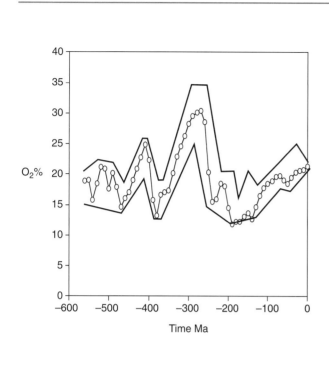

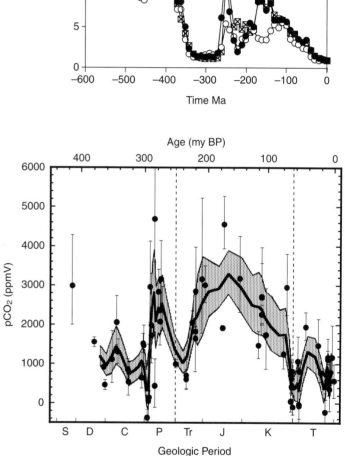

FIGURE 3.12 (Top) Phanerozoic history of O_2 and CO_2 inferred from models. SOURCE: Berner (2006). Copyright 2006 by Elsevier Science and Technology Journals. Reproduced with permission. (Right) CO_2 inferred from chemical analysis of soil carbonates. SOURCE: Ekart et al. (1999). Copyright 1999 by the *American Journal of Science.* Reproduced with permission.

fully understand. Today's atmosphere contains about 21 percent oxygen and only about 0.03 percent carbon dioxide, yet multiple lines of evidence indicate that the atmosphere contained little or no O_2 for the first 2 billion years of Earth's history (Bekker et al., 2004) and may have contained much more CO_2. In the past 500 million years, the O_2 content of the atmosphere seems to have varied from perhaps 10 to 30 percent and the CO_2 content from as low as 0.02 percent to as high as 0.7 percent (Figure 3.12). Oxygen is necessary for

many bacteria and nearly all forms of eukaryotic life and is critical to our concept of planetary habitability. Photosynthesis provides the only plausible source of this oxygen, and so oxygenation of the atmosphere and oceans constitutes an essential example of how life has profoundly influenced Earth's surface conditions. Oxygenic photosynthesis also links atmospheric oxygen with atmospheric carbon, in that the O_2 comes mostly from extracting oxygen from CO_2 and making reduced carbon in the form of organic molecules.

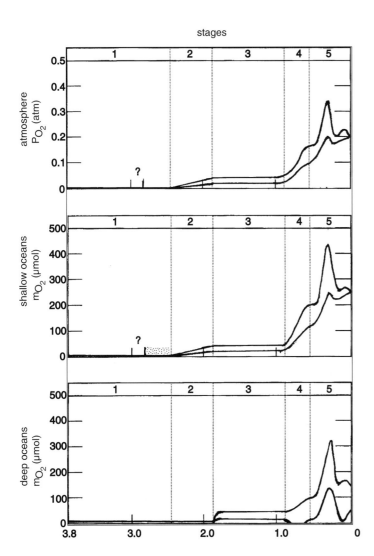

stages

FIGURE 3.13 Schematic of the rise of atmospheric O_2 concentrations. The two curves indicate the approximate range of values allowed by available data. Photosynthetic bacteria evolved no later than 2.7 Ga and perhaps as early as 3.8 Ga. Whether the diversity of early photosynthetic bacteria included the oxygen-producing cyanobacteria remains uncertain. Geological evidence indicates that whether or not oxygen-generating photosynthesis evolved in Archean oceans, O_2 did not become significant in the atmosphere and surface oceans until about 2.4 Ga. Geological evidence also suggests that there was a further delay before O_2 levels became significant in the deep ocean. Subsequently, there were times when the deep ocean became oxygen poor, even though there was appreciable oxygen in the atmosphere. SOURCE: Holland (2006). Reprinted with permission.

Oxygen began to accumulate in the atmosphere and oceans 2.3 billion to 2.45 billion years ago, but the abundance remained quite low for another 2 billion years (Figure 3.13; e.g., Brocks et al., 2005; Canfield, 2005). Considering that oxygen-generating photosynthetic bacteria were already present 2.7 billion years ago, the long delay in oxygenation of the atmosphere is hard to understand (Kopp et al., 2005). Why didn't the radiation of cyanobacteria—the only bacteria to evolve oxygenic photosynthesis and the progenitors, via endosymbiosis, of chloroplasts in algae and land plants—spread O_2 rapidly through surficial environments to produce an atmosphere like the one we have today? Part of the answer is biological: organisms that respire aerobically, from bacteria to humans, gain energy from the reaction of oxygen with organic molecules, reversing the chemistry of photosynthesis. The growth of atmospheric oxygen becomes possible when rates of oxygen production exceed those of aerobic respiration and other reactions that consume O_2. For example, burial of organic material (reduced carbon produced by photosynthesis) by sediments inhibits aerobic respiration, paving the way for oxygen accumulation in the atmosphere and oceans.

On the early Earth other processes could also have contributed to the production of molecular oxygen. For example, if the early atmosphere contained much higher amounts of both CO_2 and H_2O than it does today, substantial hydrogen could have been lost from the upper atmosphere to space. This process would have had the effect of converting H_2O to O_2. Considering the evidence that only tiny amounts of oxygen were present in Earth's early atmosphere, however, this process could not have been very efficient (Catling et al., 2001; Tian et al., 2005). Photochemical destruc-

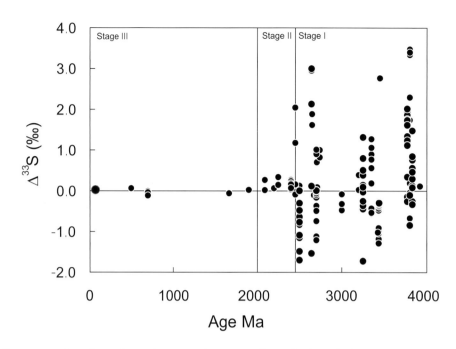

FIGURE 3.14 Sulfur isotope data that indicate the atmosphere was effectively devoid of oxygen until about 2,400 Ma. $\Delta^{33}S$ represents the mass independent sulfur isotope fractionation, which occurs at high ultraviolet radiation levels. Nonzero $\Delta^{33}S$ before 2 Ga implies that ozone (and therefore O_2), which absorbs ultraviolet radiation, had very low concentrations. SOURCE: Farquhar and Wing (2003). Copyright 2003 by Elsevier Science and Technology Journals. Reproduced with permission.

tion of methane in the Archean atmosphere could also produce hydrogen that would escape from Earth, again facilitating the oxidation of Earth's surface (Catling et al., 2001).

However oxygen accumulated in the atmosphere, its consequences were immense. Some bacteria evolved a mechanism to gain energy from the reaction of oxygen gas with organic molecules (aerobic respiration), and the ancestors of modern eukaryotes appropriated this mechanism by capturing respiring bacteria and reducing them to the metabolic slaves we know as mitochondria. It has been proposed that the bacterial ancestors of mitochondria were only facultative respirers, conducting anoxygenic (nonoxygen-producing) photosynthesis in oxygen-free environments (Woese, 1977). If true, the original basis of ecological interaction could have been photosynthetic—but its lasting legacy was unquestionably aerobic respiration in nucleated cells.

We do not yet understand how biological, tectonic, volcanic, and atmospheric processes combined to produce the episodic rise in the amount of oxygen in the atmosphere. In fact, we are only now developing the analytical tools needed to read Earth's long-term

environmental record at high resolution. For example, the recent discovery that the isotopic composition of atmospheric sulfur was subtly different before 2.5 billion years ago (Figure 3.14) confirms that the concentration of O_2 in the atmosphere back then must have been less than 10^{-5} of the present level (Farquhar et al., 2000; Pavlov and Kasting, 2002)—effectively oxygen free.

The rise of atmospheric oxygen also eventually produced a rise in the level of atmospheric ozone, which shields Earth's surface from ultraviolet radiation that is detrimental to life on land. The ozone concentration that is sufficient to provide full ultraviolet shielding is surprisingly small—an atmospheric O_2 concentration about 1 percent of the modern level (Kasting et al., 1985), a level that was probably reached soon after 2.5 billion years ago.

We do not understand whether the initial evolution of cyanobacteria triggered the first round of oxygen accumulation or preceded it by hundreds of millions of years; nor do we understand why oxygen levels remained low through most of the Proterozoic Eon (2,500 to 542 million years ago) or what processes drove the renewed increase that paved the way for animal diversification. For that matter, we do not understand why the modern atmosphere contains the amount of oxygen it does. Oxygen-related questions are sufficiently complex that they will require expanded research interactions among Earth scientists, atmospheric scientists, and biologists. For example, we need better paleontological resolution of when oxygenic photosynthesis first evolved and when the eukaryotic organisms that now dominate primary production rose to global prominence (Falkowski et al., 2004). There is still no reliable geochemical proxy for ancient oxygen abundances (Berner et al., 2003), and models relating deep-Earth processes to surface conditions do not yet take account of historical patterns and feedbacks from physiology, tectonics, and atmospheric chemistry.

Major questions of oxygen history are not limited to its long-term trajectory. During the Paleozoic and Mesozoic eras, wide portions of the oceans beneath the surface mixed layer became essentially "oxygen deserts," a condition known as anoxia. Geologically transient but globally distributed oceanic anoxic events are well documented from Early Jurassic, Early Cretaceous, and Late Cretaceous rocks (Jenkyns, 2003). In modern oceans, O_2 can fall to low levels and, locally, may decline to zero in the ocean-minimum zone just below the well-mixed surface water mass in which most photosynthesis takes place. What makes the oceanic anoxic events stand out is the large spatial scale of anoxic water masses. So far, we know that these events coincide with perturbations in the carbon cycle, as deduced from records of the isotopic composition of marine carbon and the strontium isotopic composition of seawater (e.g., Jones and Jenkyns, 2001; see Question 7). But we do not know what tipped the redox balance, causing anoxia to spread repeatedly through Mesozoic oceans; nor do we know why there are similar (but less well documented) events in Paleozoic oceans but none from the Cenozoic Era. And we do not know whether these events were produced entirely by inorganic geological processes or whether organisms exacerbated, ameliorated, or otherwise responded to these events.

As oxygen levels increased in the atmosphere and oceans, new forms of life became possible. Animals that move about in search of food have an elevated need for oxygen, and so it is not surprising that the first evidence for large animals with high oxygen demands met (initially) by diffusion through tissues

coincides with geochemical evidence for elevated O_2. Oxygen levels may have reached historically high levels (perhaps as much as 30 percent of the atmosphere, by volume) some 300 Ma, potentially explaining how pigeon-sized dragonflies could fly above tropical forests of the day (Dudley, 2000). Sharp, if transient, depletion of oxygen in ocean waters had the opposite effect, reducing animal diversity and size in widespread areas of the seafloor—a particularly widespread episode of marine anoxia is associated with mass extinction at the Permian-Triassic boundary, some 250 Ma (Wignall and Twitchett, 1996).

Other Interactions Between Earth and Life

Oxygen provides a compelling example of rapidly unfolding research on the interactions between the physical and biological Earth, but it is hardly the only example. Carbon dioxide is also intimately related to biological activity, not only through climate and the carbon cycle (Royer et al., 2001) but also because it affects the ability of marine organisms to form carbonate skeletons (Kleypas et al., 2006). The physiological link between skeletons and carbon dioxide may help explain some major biological changes of the past. For example, accelerating physiological research on the biological consequences of ocean acidification illuminates Earth's greatest mass extinction at the end of the Permian period (252 Ma) when marine ecosystems collapsed (e.g., see below). Similarly, current increases in carbon dioxide raise concern about the future of reef corals and other organisms that form carbonate skeletons in the shallow ocean. The other side of this interaction—how biological and physical processes interact to govern CO_2 on both short and long timescales—is also a major issue in Earth history, one of immense importance as we debate the consequences of current human activities (Question 7). As in the case of oxygen, deeper understanding of the feedbacks between life and carbon dioxide levels, on scales from the local and ephemeral to those governing the long-term history of the planet, will require better geochemical and paleobiological proxies for ancient CO_2 abundances, more nuanced understanding of the biological processes that influence carbon dioxide levels (especially those related to microorganisms and plants), and increasingly sophisticated models that account for both biological and physical parameters.

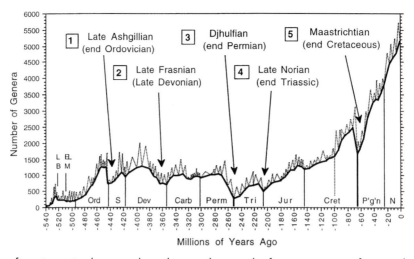

FIGURE 3.15 Number of marine animal genera through time, showing the five major times of "mass depletion" in biological diversity. Only three drops—end-Ordovician, end-Permian, and Cretaceous-Tertiary—are driven primarily by increases in extinction rates, rather than declines in rate of origin. SOURCE: Bambach et al. (2004). Copyright 2004 by the Paleontological Society, Inc. Reproduced with permission.

Problems as diverse as the influence of rainforests on Earth's hydrological cycle, the role of vegetation in stabilizing the land surface, the relationship between nutrient availability and diversity, and the oceanwide biogeochemical consequences of deep-water anoxia engage a wide range of Earth scientists because they have both deep-time evolutionary components—how did the diversification of woody plants change Earth's surface?—and topical applications—what will be the consequences for the Earth system of rainforest clear-cutting, increased soil erosion, and seafloor anoxia linked to fertilizer-spiked nutrient flows from agricultural lands to the ocean? Earth scientists have almost limitless opportunities to join with biologists to fashion both a new picture of our planet's history and a clearer picture of our future.

What Caused Mass Extinctions?

Nothing illustrates how heavily life depends on a favorable surface environment as clearly as a sharp change in that environment—which has occurred several times during the past 500 million years, causing the mass extinction of species (Figure 3.15). In particular, the great extinctions at the end of the Permian (252 Ma) and Cretaceous (65 Ma) periods influenced the course of biological evolution as much as all the accumulated genetic changes during the 187 million years between

them. But what specific events or environmental changes precipitated the great mass extinctions, and what aspects of biology influenced the patterns of survival and recovery, are not known. Most Earth scientists agree that a meteorite impact caused the end-Cretaceous extinction of dinosaurs, ammonites, and myriad other plant, animal, and microscopic species (Alvarez, 1997), but the actual kill mechanisms unleashed by this trigger remain poorly understood. The relative importance of coincident environmental perturbations, including an interval of oceanographically driven global change, extensive extrusion of flood basalts, and the particular location of the impact on a tropical continental platform, are simply not known.

Although a single plausible event may account for the end-Cretaceous extinctions, the cause of the end-Permian mass extinction, which may have erased as many as 90 percent of marine species and many terrestrial species (Erwin, 2006), is still debated. Support for an extraterrestrial cause is limited, with growing interest in direct and indirect effects of massive volcanism in what is now Siberia. An emerging view is that massive flood basalts, intruded through thick carbonates and extruded onto thick peat deposits, produced unusually high emissions of carbon dioxide and thermogenic methane, resulting in global warming, acidification of the oceans, depletion of oxygen in ocean waters below the mixed layer, and enhanced production of hydrogen sulfide by bacteria living in those oxygen-depleted water masses. Physiological research on modern marine organisms, aimed at understanding current environmental change (e.g., Pörtner et al., 2005), allows Earth scientists to predict the biological consequences of such an event on end-Permian biological diversity. Indeed, paleobiological data show that extinctions did not affect all Permian animals equally. For example, groups whose living relatives were vulnerable to the physiological consequences of sea acidification disappeared at rates much higher than those physiologically well buffered against such environmental perturbations. Extinctions on land are consistent with the predicted effects of rapid climate change (summarized in Knoll et al., 2007). Continuing research on Earth's great intervals of biological upheaval will increasingly integrate insights from paleobiology, stratigraphy, high-precision geochronology, and geochemistry with physiology and models generated to help understand current issues of global change.

What Governs the History of Biological Diversity?

Major extinctions have clearly influenced the history of plant and animal life, but what, fundamentally, controls the observed pattern of diversity increase from the Cambrian to today (Figure 3.15)? Quantification of diversity change through time on land and in the oceans remains a subject of active research and debate, but many Earth scientists would agree that the modern world (at least in preindustrial times) harbors more species of land plants, more species of land animals, and more species of marine animals than any previous moment in our planet's history (e.g., Benton and Emerson, 2007). Attempts to model diversity history employ logistic equations, which imply biologically or physically imposed limits to diversification (e.g., Sepkoski, 1984), or exponential equations, which imply persistent diversity increases, episodically knocked back by mass extinctions (Stanley, 2007).

The tension between these classes of models focuses attention on a great and unsolved problem. What are the relative roles of genetic innovation, ecology, and physical Earth history in governing the long-term history of life? The answer certainly requires macroecological insights from biologists, but the questions are necessarily framed by paleontologists. And

rapidly emerging insights into the physical history of the Earth surface system provide, for the first time, the proper environmental framework to address the issue. Has primary production increased through time, and if so what have been its consequences? What are the consequences of sea-level change, episodically flooding and exposing continental interiors, on species origination and extinction in the marine realm (Peters, 2005)? Did the rules of community construction change when flowering plants evolved the capacity to use animals to ensure the faithful spread of pollen from one plant to the next? How did the ecological relationships that undergird community diversity reform following episodes of mass extinction? Detailed analyses of community organization in systems as disparate as Pleistocene coral reefs, Cenozoic mammals, and Carboniferous forests promise important insights into ecology and evolution that cannot be made solely on the basis of the short-term observations and experiments available to biologists (Jackson and Erwin, 2006).

Summary

Earth's surface environment is obviously altered by large-scale geological processes (Questions 4 and 5), but it is also affected continuously and pervasively by the activities of life forms. Likewise, Earth's geological evolution and infrequent catastrophic events, such as meteorite impacts, have clearly affected the evolution of life. But even when we can document extinctions and major evolutionary changes, we cannot yet sort out the causes. To what extent were they caused by geological as opposed to biological processes? Which environmental conditions were responsible for which extinctions or changes in biological form and function? We know that the composition of Earth's atmosphere, especially its high concentration of oxygen, is a major consequence of the presence of life, one that made possible the evolution of more complex organisms. But exactly how other geological events have affected evolution, and how much control life has had on climate, are still topics of debate.

Life processes and Earth processes also interact locally. Erosion rates, climate, and weathering rates affect the habitability of specific regions of Earth, and the ecosystems themselves in turn affect erosion rates, climate, and weathering processes. Understanding the interrelationships between surficial processes that shape the land and the life that inhabits it presents a critical challenge for managing land resources and becomes even more important as we attempt to forecast the effects of future climate change.

Understanding how Earth's life and geological environment arrived at their present state, and how they interacted in doing so, constitutes a major intellectual challenge. Meeting this challenge will help us understand how life will respond to present-day environmental change, but Earth scientists will have to develop new research and educational partnerships with biologists and atmospheric scientists. The search for life on extrasolar planets will similarly depend on better understanding of biogeochemical influences on atmospheric composition here at home.

4

Hazards and Resources

Geological processes affect the sustenance and safety of the human population. For example, earthquakes and volcanic eruptions can cause widespread damage and loss of life. Mineral and water resources are necessary to maintain our complex societies, and waste products from resource use must somehow be returned to Earth in a manner that does not unnecessarily foul the environment and that can be sustained for the indefinite future. The catastrophic nature of some Earth processes and the wise use of our land, water, and mineral resources present a special set of challenging research issues. It is likely that Earth will remain habitable to humans for millions of years into the future, barring insurmountable environmental degradation or a catastrophe of the type that struck 65 million years ago. This last chapter deals with fundamental Earth science that must be advanced to ensure and enhance the future of humankind. We are most concerned as a society about the next decades and centuries, but it is interesting, instructive, and scientifically challenging to think much farther ahead (and to look back in deep geological time) to fully comprehend the range of possibilities.

This chapter comprises two grand questions. Question 9 addresses earthquakes and volcanic eruptions, essential planetary processes that sometimes are deleterious to humankind but are intrinsic to Earth and probably inseparable from its habitability. Both of these geological phenomena are catastrophic in the sense that they represent the sudden release of energy stored inside Earth. The best approach to minimize the loss of life and property from these events is to understand both the causative processes well enough to forecast or predict them and their effects well enough to address them. Question 10 is aimed at addressing the fundamental science that underlies many of the issues related to land, mineral, and water resource use, as well as waste disposal. We have focused this question on the science of Earth fluids, which arguably represents the single most central, fundamental, and crosscutting research area for environmental management and sustainability.

QUESTION 9: CAN EARTHQUAKES, VOLCANIC ERUPTIONS, AND THEIR CONSEQUENCES BE PREDICTED?

Earthquakes and volcanic eruptions are sudden and hazardous manifestations of the normally gradual movements of Earth's interior (Questions 4 and 5). Although much research is stimulated by the dangers these events pose to human populations, they are of special interest to Earth scientists for other reasons as well. They constitute a class of phenomena that can be observed in action and monitored at many different scales and that recur frequently. The fact that we know they will happen, and also (for the most part) where, creates a natural desire to be able to predict them. The imperative for improved predictive power is escalated as human populations increasingly concentrate in areas prone to earthquakes and volcanic eruptions. But prediction remains difficult because of the inherent complexity of the processes and the special demands imposed by attempting to specify exactly when these events will occur. And since even highly accurate predictions will not prevent widespread damage, an improved ability to forecast the consequences of catastrophic events, and hence to prepare for them, is at least as important as predicting them.

Earthquake Hazards

Earthquakes at their worst are extreme catastrophes. The 1556 Shaanxi, China, earthquake killed over 800,000 people in a matter of minutes. By some estimates the next large earthquake under Tokyo could cause trillions of dollars in direct economic losses. These consequences could be mitigated if earthquakes could be predicted over timescales short enough to allow an effective response. However, this goal remains elusive.

The nature of earthquakes makes them uniquely terrifying. No one sees an earthquake coming. It is a matter of seconds from the time shaking first becomes perceptible until it becomes violent. At any locale, damaging earthquakes occur infrequently on human timescales, which means that most people caught in a major earthquake have no previous experience. It is also profoundly disturbing when our usually stable reference frame, the planet beneath us, does not hold still. The unpredictability and sudden onset of earthquakes also mean that once an earthquake begins, it is generally too late to do much more than duck and cover. The combination of unpredictability, abrupt onset, rarity, and unfamiliarity means that the risk posed by earthquakes is difficult to manage, for both individuals and governments.

Earthquake Prediction: Where, When, and How Big?

The goal of earthquake prediction is to specify where and when a significant earthquake will occur. Where future earthquakes will occur is largely understood, with some important exceptions (summarized in Beroza and Kanamori, 2007). Predicting when they will strike is much more difficult, though progress has been made and promising avenues of research have emerged. The term "significant" is a subtle but important part of the definition of earthquake prediction, and it brings up the important question of what controls earthquake size.

Finally, even the word "earthquake" needs definition. Scientists use the term to describe the fault rupture that generates seismic waves. However, the public views an earthquake more broadly as both the faulting and the waves. We will use this more general definition and discuss prediction of the faulting event and the shaking that accompanies it.

Predicting Where Earthquakes Happen

Scientists have long recognized that some regions are seismically active, while others are not. By the middle of the 20th century, seismologists had produced a remarkably complete earthquake atlas (Gutenberg and Richter, 1954) that chronicled systematic features of global seismicity, but they lacked a framework to understand those features. The advent of plate tectonics soon changed that and also enabled the first steps toward earthquake prediction to be taken. For example, plate tectonics theory led to the recognition that Cascadia should be subject to large earthquakes, despite no history of earthquake activity. This expectation was confirmed by the discovery, using stratigraphic and other evidence, of a magnitude (M) ~ 9 earthquake in Cascadia in January 1700 (Figure 4.1; Atwater, 1987).

Plate tectonics holds that Earth's lithosphere consists of large plates that move relative to one another at speeds of several centimeters per year (Question 5). Relative plate motion is accommodated on plate-boundary faults or, more typically, on complex fault systems. These faults are frictionally locked between earthquakes, causing ongoing plate motion to deform the crust around them, storing elastic strain energy in the process. Once friction is overcome and the fault starts slipping in an earthquake, this stored energy is converted to other forms, most notably energy radiated away from the fault as seismic waves.

Most earthquakes occur at plate boundaries, and the type of boundary plays a role in controlling the nature of earthquake activity. Extension at divergent boundaries is accommodated by normal faulting and formation of new crust through basaltic volcanism, with much of the deformation taking place aseismically. Horizontal motion across transcurrent plate boundaries takes place on strike-slip faults. Most transcurrent boundaries are oceanic, but when they traverse continents, they pose significant seismic hazard. All of the largest earth-

FIGURE 4.1 A "ghost forest" near the mouth of the Copalis River, Washington, that was killed by saltwater tides after a M ~ 9 earthquake in January 1700 caused the land to subside. SOURCE: <http://soundwaves.usgs.gov/2005/07/outreach.html>. See also Atwater et al. (2005).

quakes and most of Earth's seismicity occur at convergent plate boundaries. The subduction of relatively cool oceanic crust increases the depth of the elastic-brittle regime in which earthquakes occur. Moreover, when the subducted slab crosses the elastic-brittle regime at a shallow angle (e.g., in Sumatra, Alaska, and Chile), the seismogenic zone can be hundreds of kilometers wide. The hazard at convergent boundaries is not always commensurate with earthquake size because much of the faulting and the strongest shaking occur underwater. However, the 2004 tsunami (Figure 4.2) illustrates the destructive potential and global reach of large subduction zone earthquakes.

Earthquakes that occur within a tectonic plate account for less than 1 percent of the world's earthquakes, but they pose a significant seismic hazard and can be quite large. For example, a sequence of strong earthquakes with magnitudes as high as 8 shook New Madrid, Missouri, for eight weeks in 1811-1812, destroying the town and causing widespread destruction across the central United States. Intraplate earthquakes are not readily explained by plate tectonics. Some occur within broad plate boundary deformation zones, such

as those across south Asia and western North America, while others are not clearly associated with any plate boundary, such as those in Australia or eastern North America. Half of all intraplate earthquakes occur in failed continental rifts (Johnston and Kanter, 1990), but their underlying cause remains a mystery. Putative explanations for intraplate earthquakes include localized stresses induced by emplacement and crystallization of magma below the surface, postglacial rebound, and weak zones in otherwise strong crust.

Intermediate (70 to 300 km deep) and deep (300 to 700 km deep) focus earthquakes occur at convergent plate boundaries within subducting lithosphere (Figure 2.10). Although they pose less of a threat than shallow, plate-boundary earthquakes, intermediate-depth earthquakes can be quite destructive (Beck et al., 1998). What causes them is unclear because Earth materials are expected to deform plastically at the depths where they occur (Question 4). Candidate mechanisms to explain intermediate and deep earthquakes include elevated fluid pressures, accelerating deformation and frictional heating, and mineral phase changes (Kirby et al., 1996).

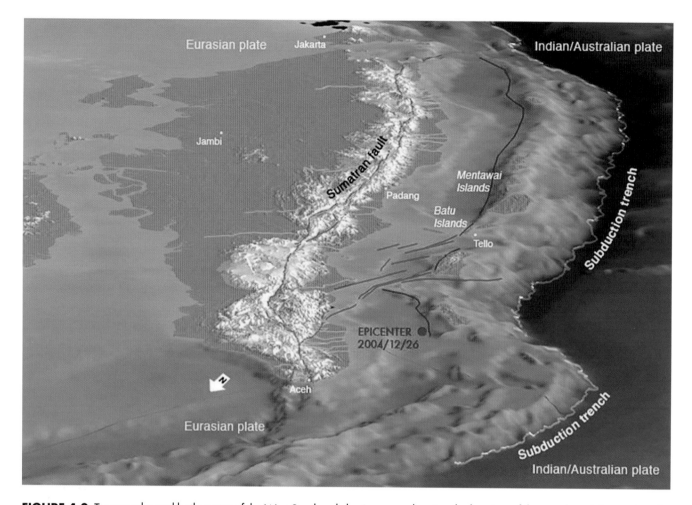

FIGURE 4.2 Topography and bathymetry of the West Sunda subduction zone showing the location of the trench and the earthquake fracture zones (colored lines). The red dot shows the epicenter of the M 9.1 December 2004 earthquake that ruptured the India/Eurasia plate boundary (area between the trench and Sumatra) and caused the devastating Indian Ocean tsunami. Rupture propagation was primarily northward, toward the lower left of the figure, extending hundreds of kilometers beyond the figure. Subsequent earthquakes have ruptured parts of the plate boundary to the south. SOURCE: Courtesy of Mohamed Chlieh, Caltech. See also Chlieh et al. (2007) and <http://today.caltech.edu/gps.sieh/>. Used with permission.

Predicting When Earthquakes Will Happen

Just as plate tectonics explains where most earthquakes occur, it has much to say about how often they occur. The velocity of relative motion across plate boundaries is known to within several millimeters per year, which provides a boundary condition on fault-slip rates, and thus on how frequently earthquakes must occur over the long term. The utility of this boundary condition for earthquake prediction is confounded by the fact that plate boundaries typically comprise complex fault systems, and the partitioning of slip among faults is difficult to unravel, even in well-studied systems such as the San Andreas. In some cases, slip rates are well

known, but even then the irregular recurrence of earthquakes makes forecasting difficult.

Earthquake predictions are commonly classified by time frame. The types of predictions discussed below are:

1. *Long-term forecasts of events of an uncertain magnitude that have a low probability of occurrence over a large window of time.* Long-term forecasts based on probabilistic methods are an active area of research.

2. *Short-term prediction of events of a specific size that have a high probability of occurrence within a narrow range of space and time, weeks or months in advance.* There is currently no way to predict the days or months when

an event will occur in any specific location, and it is not clear whether it will ever be possible.

3. *Early warning that seismic waves from a developing event will arrive in seconds, ideally enabling an alert to be issued before damaging strong ground motions begin.* This emerging area of research shows promise for reducing seismic risk.

Long-Term Forecasting

Earthquakes can be forecast by assuming that they occur randomly in time but at known long-term rates. If no historical record of large earthquakes exists, the long-term rate is extrapolated from the frequency-magnitude relationship of small earthquakes. Unfortunately, the assumption that earthquakes occur randomly in time is highly suspect. Earthquakes are observed to cluster in both space and time. Moreover, the elastic-rebound hypothesis suggests that once a large earthquake releases the accumulated stress on a fault, that same fault segment is unlikely to experience a large earthquake until strain has reaccumulated. Such considerations have led to time-dependent earthquake forecasts.

Perhaps the simplest time-dependent, long-term forecast is that offered by the seismic gap hypothesis, which posits that a future earthquake is more likely on a part of a fault that has not ruptured recently (Fedotov, 1965). Implicit in this hypothesis is the notion of a "characteristic earthquake," that is, a large earthquake that defines a fault segment and dominates the slip budget. According to the seismic gap hypothesis, the probability of an earthquake will be small immediately after the previous earthquake, and the conditional probability that a characteristic earthquake will occur can be determined from the time of the previous earthquake. The evidence for characteristic earthquake behavior is equivocal, and tests of the seismic gap theory have called its utility into question (Kagan and Jackson, 1991). However, the notion that a fault must keep up with geological slip rates over the long term seems inescapable, so there ought to be information in the system that can inform earthquake forecasts. Developing accurate forecast models that use this information is an area of ongoing research.

A number of time-dependent models take some account of how fault systems are thought to operate. Because they need to draw on diverse aspects of fault system behavior, long-term earthquake forecast models form a framework for integrative research. Paleoseismology—the investigation of individual earthquakes in the geological record—provides critical information documenting the frequency and variability of earthquake occurrence over the long term. Geodetic measurements (e.g., Figure 4.3) constrain the distribution of strain accumulation. Earthquake source models constrain the amount of slip expected in large earthquakes. Computer algorithms that scan seismicity catalogs and account for fault slip and interaction, in concert with historical earthquake catalogs, constrain spatial and temporal earthquake probabilities. Finally, models of static and dynamic triggering help us understand how earthquakes interact, including how the probability of one earthquake increases or lessens the probability of another (see below). A longstanding and difficult problem that would benefit from further research is how we validate long-term earthquake forecasts.

The discussion above concerns shallow earthquakes for which we understand the factors influencing the frequency of recurrence, such as fault-slip rates that must be satisfied over the long term. For both intermediate and deep-focus earthquakes, we lack the kinematic fault-slip boundary conditions that enable us to constrain the long-term probabilities of shallow earthquakes. Until a better understanding of intermediate and deep-focus earthquake recurrence is achieved, long-term forecasting of such events will remain empirical.

Are Short-Term Predictions Possible?

The challenge of short-term earthquake prediction can be illustrated by drawing an analogy with lightning. Both phenomena involve the abrupt conversion of accumulated potential energy to kinetic energy. In the case of lightning, gradually accumulated electrical charge suddenly flows as electric current in a lightning bolt and radiates sound waves as thunder. For earthquakes it is gradually accumulated elastic strain energy that accelerates the crust on both sides of the fault and radiates energy as seismic waves. Based on the relative timescales, predicting the size, location, and time of an earthquake to within a week corresponds to predicting the size, location, and time of a lightning bolt to within a millisecond. The latter sounds hopeless but would ac-

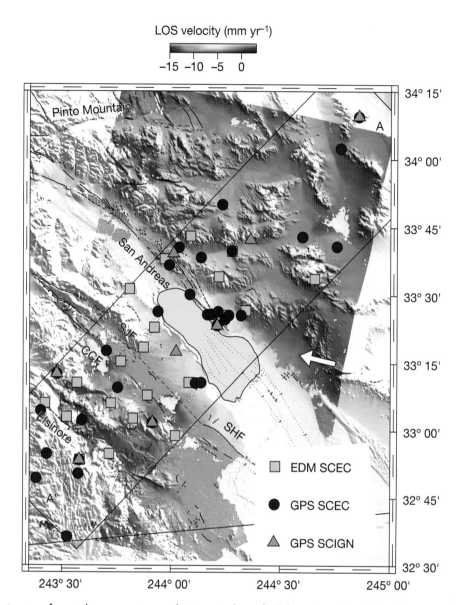

FIGURE 4.3 Estimates of crustal movement near the San Andreas fault based on decadal-scale measurements. Color combines information from Global Positioning System (GPS) measurements and spaceborne radar interferometry (InSAR). Scale at the top shows velocity in the satellite line of sight as it passes over the region. The blue region in the lower left, southwest of the San Andreas and San Jacinto (SJF) faults, is moving away from the satellite track at ~ 14 mm/yr, which corresponds to northwestward motion at ~ 45 mm/yr relative to the region in red to the northeast of the faults. This suggests that strain is accumulating on the San Andreas fault in this region, where no large earthquake has occurred in over 250 years. (CCF = Coyote Creek fault, EDM = Electro-optical Distance Measurement, SCEC = Southern California Earthquake Center, SCIGN = Southern California Integrated GPS Network, SHF = Superstition Hills fault). SOURCE: Fialko (2006). Reprinted by permission from Macmillan Publishers, Ltd.: *Nature*, copyright 2006.

tually be straightforward because lightning is preceded by an easily observable nucleation process—the formation of a channel of ionized air, known as a stepped leader, that provides a conductive path for the current in the main lightning bolt. The stepped leader precedes the lightning strike over its entire length. Prediction of the time and location of a lightning bolt to within a

millisecond is easy once the stepped leader forms. Do earthquakes have a similar nucleation process?

Laboratory studies (e.g., Dieterich, 1979) and theoretical models (e.g., Andrews, 1976) of earthquake nucleation indicate that unstable fault slip should be preceded by an aseismic nucleation process. It seems likely that nucleation of some sort must occur before earthquakes, but if so, how extensive is the nucleation process? If it occurs over a limited part of the fault, and thereafter earthquake rupture becomes self-sustaining, then earthquake prediction will be practically impossible. If, on the other hand, nucleation scales with the size of the eventual earthquake (e.g., if the size of the nucleation zone is proportional to the size of the eventual earthquake), prediction would be a good deal more likely, though still extremely challenging (Ellsworth and Beroza, 1995). The nature of earthquake nucleation is a key unknown that is central to the question of short-term earthquake predictability.

Another possible way to predict earthquakes in the short term is through patterns of earthquake interaction. Large shallow earthquakes are immediately followed by aftershocks that are triggered by the main shock. Large earthquakes sometimes trigger other large earthquakes, most famously in the sequence of large earthquakes that ruptured most of the North Anatolian fault during the 20th century (Toksöz et al., 1979). Thus, the study of aftershocks provides insight on how large earthquakes might interact and on how small earthquakes might trigger large earthquakes.

Aftershocks reflect the response to a stress change imposed by the main shock (Scholz, 1990), but this is not a complete explanation because static, elastic effects by themselves cannot explain the observed gradual decay of aftershock rates. The mechanisms proposed to explain aftershock decay include pore fluid flow, viscoelastic relaxation, and earthquake nucleation under rate- and state-variable friction. A better understanding of these mechanisms would improve prediction of earthquake interaction.

Less obvious than the triggering of earthquakes by static stress changes is their suppression when the stress necessary for earthquakes is relieved by a nearby earthquake. A well-known example of this stress shadow effect is the nearly complete absence of major earthquakes in northern California following the 1906 San Francisco earthquake (Ellsworth et al., 1981). Dramatically fewer earthquakes have occurred in this region in the 100+ years since that earthquake than occurred in the 50 years leading up to it. An obvious conclusion is that the 1906 earthquake suppressed subsequent earthquakes by relieving the shear stress in Earth's crust.

Following the 1992 Landers earthquake, small earthquakes occurred more than 1,000 km away (Hill et al., 1993). Static stress changes over such distances are negligible, but dynamic stress changes transmitted by seismic waves have been documented in the 1999 Izmit, 2002 Denali, and 2004 Sumatra earthquakes. In the case of Sumatra, earthquakes were triggered in Alaska, a distance of 11,000 miles (West et al., 2005). The Landers trigger was synchronous with the maximum vertical displacement of large and extremely long-period surface waves, indicating a direct role for dynamic stresses. This is a new concept of earthquake interaction, and dynamic stress changes can presumably trigger earthquakes at short distances as well.

Whether by static triggering, dynamic triggering, or other mechanisms, the occurrence of an earthquake affects the probability of future earthquakes nearby. In aftershock sequences, triggered earthquakes can themselves trigger other earthquakes in a cascade of failures. Models that quantify these so-called epidemic-type aftershock sequence interactions form the basis for a variety of probabilistic short-term earthquake forecasts. Improving these forecasts and testing their skill at prediction are areas of active research (Field, 2007).

Complexity

Earthquake processes span a tremendous range of temporal and spatial scales, which makes them intrinsically difficult to characterize, let alone predict. Spatial scales range from the size of individual mineral grains to the size of tectonic plates. The smallest microearthquakes rupture faults for milliseconds, whereas strain accumulation during the earthquake cycle can be thousands of years. Physical mechanisms that are dominant at one scale might become negligible at others. Superimposed on the scale variations is the complexity of geological structures and materials.

Earthquakes and fault systems have been held up as an example of a complex natural system that exhibits self-organized criticality (Bak et al., 1988). Despite the complexity, earthquake phenomena exhibit certain types order. Earthquake stress drop and radiation efficiency are similar for both large and small earthquakes. Gutenberg-Richter statistics, a power-law description of the relative number of large and small earthquakes, appear to apply for all earthquake populations. Omori's Law provides a universal description of the rate of aftershock decay. The geometry of fault networks is typically treated using a fractal description. Methods of statistical physics are used to understand how these relationships emerge from the earthquake process and to predict their behavior (e.g., Turcotte et al., 2007). If fault systems behave chaotically, as suggested by some models, there may be an intrinsic limit to predictability (NRC, 2003b). This limit might be years or months if we could fill gaps in our knowledge of the physical laws governing fault motion and if it were possible to measure accurately all the stresses and strains in and around the fault.

How Much Warning Can Be Given Before an Earthquake?

Real-time seismology, which became possible with the regional deployment of high-quality instrumentation and rapid, continuous telemetry, provides reliable estimates of the location and size of earthquakes within a few minutes of the initiation of rupture. For nearby earthquakes, ground shaking will have already begun before these estimates can be made. Earthquake early warning systems focus on the seconds after an earthquake rupture has already started. These systems exploit the fact that the speed of telecommunications exceeds that of seismic waves. If seismographs can quickly determine that an earthquake is under way, and importantly that it is a large earthquake, then regions likely to be subject to dangerous shaking can be alerted before the seismic waves arrive. Earthquake early warning systems are operational in Japan (Figure 4.4), Mexico, and Taiwan and are in various stages of development in Romania, Turkey, and the United States. The key to earthquake warning systems is rapid determination that a large earthquake is under way before the earthquake has fully developed (Allen and Kanamori, 2003). The extent to which this is possible is closely tied to the nature of earthquake nucleation discussed above. The amount of warning that earthquake early warning systems can provide for large earthquakes can be tens of seconds under favorable circumstances.

Strong Ground Motion Prediction

Predicting the level of damaging shaking from seismic waves in an earthquake is a critical aspect of both earthquake prediction and risk mitigation. But even if short-term earthquake prediction ever became a reality, it would still be impossible to protect most of the built environment from damage. Predicting strong ground motion is itself a considerable scientific challenge. Earth's crust is strongly heterogeneous at all scales, so earthquake waves are strongly distorted as they propagate through it. The faulting process is also complex and may represent the dominant source of uncertainty in strong ground motion prediction. Predictions of strong ground motions are generally made using probabilistic methods and computer simulations.

Probabilistic seismic hazard analysis. The probability of strong ground motions is commonly calculated using probabilistic seismic hazard analysis. The analysis may yield, for example, an estimate of ground motion intensity that has a 2 percent probability of being exceeded over a 50-year time interval. Probabilistic seismic hazard analysis combines information on earthquake likelihoods from long-term forecasts and data on peak ground acceleration, spectral acceleration, and peak ground velocity to create a map of intensities at the specified exceedence probability. Such maps can be used to develop design criteria for buildings and to set priorities among risk reduction measures. As with other earthquake prediction tools, it will be difficult to test the validity of these predictions until the instrumental record is considerably longer. However, observations of precariously balanced rocks (Figure 4.5) have recently been used to place bounds on maximum exceeded ground motion amplitudes over time intervals as long as thousands of years (Brune, 1996).

Ground motion prediction through simulation. Very few recordings exist of strong ground motion close to large earthquakes. This is unfortunate because large earthquakes often dominate seismic hazard. Computer simulation of strong ground motion provides a possible means to fill this data gap, as long as we can be confident the simulations are accurate. Major sources of uncertainty in these calculations include characterization of the earthquake source (Figure 4.6), the ability to model the effects of wave propagation through Earth's crust, changes to the wavefield due to near-surface nonlinearity, and earthquake-to-earthquake variability in the rupture characteristics. Physics-based predic-

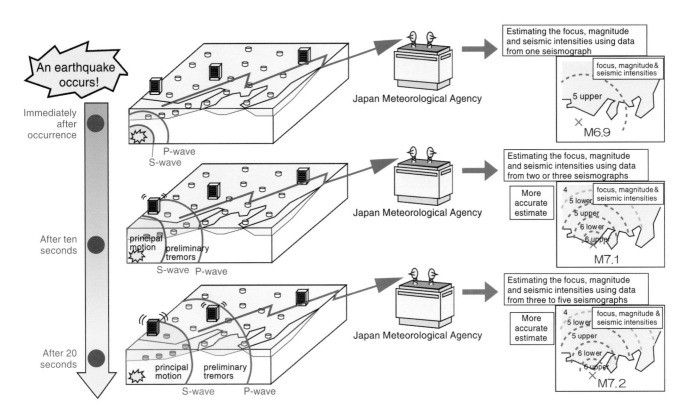

FIGURE 4.4 Schematic representation of the Japanese earthquake early warning system developed by the Japan Meteorological Agency. This system calculates the location and magnitude of the earthquake from recordings near the epicenter and then estimates the distribution of shaking more widely before the arrival of the strongest shaking, which is typically comprised of S waves. Earthquake early warnings will be broadcast through media outlets such as TV and radio. The system went public in October 2007. SOURCE: Courtesy of the Japan Meteorological Agency. <http://www.jma.go.jp/jma/en/Activities/eew1.html>. Used with permission.

FIGURE 4.5 The ground motion intensity thresholds at which precariously balanced rocks, such as the one shown at left, would be toppled provide an important test on ground motion exceedence probabilities determined from probabilistic seismic hazard analysis. SOURCE: <http://www.seismo.unr.edu/PrecRock/DSC00335.JPG>. Used with permission.

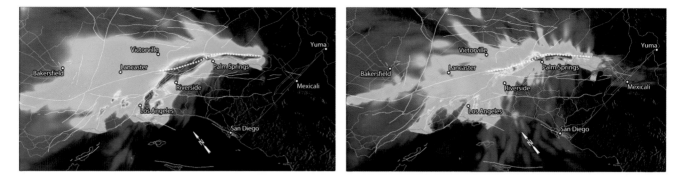

FIGURE 4.6 Ground motion intensities (warm colors correspond to high intensities) for a simulated M 7.7 earthquake with southeast to northwest rupture on 200-km section of the San Andreas fault. There is a strong rupture directivity effect and strong amplification due to funneling of seismic waves through sedimentary basins south of the San Bernardino and San Gabriel mountains. The simulation on the left assumes a kinematic rupture model, and the one on the right assumes a dynamic (physics-based) rupture model. The extreme difference in the predicted intensities underscores the importance of properly characterizing earthquake source processes. SOURCE: <http://visservices.sdsc.edu/projects/scec/terashake/compare/>. Visualization courtesy of Amit Chourasia, San Diego Supercomputer Center, based on data provided by Kim Olsen and colleagues, Southern California Earthquake Center. Used with permission.

tion of strong ground motions through simulations is an area of intense research. Creating simulations that reach high enough frequencies for structural engineering purposes requires high-performance computing. Further improvements, particularly in the validation of simulation results, are required before they will have an impact on engineering practice.

Dynamic rupture modeling. The evolution of rupture on faults can be modeled either in terms of the displacements or as a function of the stresses. The former, the so-called kinematic description, is most common, but the latter "dynamic" approach provides a more complete description of the process of fault failure and hence is an ongoing research focus. In dynamic rupture models the redistribution of stored strain energy leads to shear failure that becomes unstable and self-sustaining—the process that is believed to occur in an earthquake. If the assumptions that go into them are correct, dynamic models can serve as a foundation for better predictions of both fault behavior and strong ground motion (Figure 4.6). However, the models are computationally intensive and require an understanding of fault behavior over a wide range of conditions (e.g., slip, slip-rate, temperature, pressure, pore pressure) and physical mechanisms (e.g., slip-weakening, rate- and state-variable friction, thermal pressurization, flash heating).

What Is the Role of Slow Earthquakes?

Over the past decade seismologists and geodesists have discovered an entirely new family of unusual earthquakes that range in size from M 1 to at least M 7.5. They occur in diverse geological environments—from the subduction zones of Japan, Mexico, Cascadia, and Alaska, to the slopes of Kilauea volcano in Hawaii, to the San Andreas fault in California. They appear to be caused by the same mechanism as ordinary earthquakes but take such a long time to happen that they are described as "slow." Because these earthquakes are slow, the waves they generate, if they generate waves at all, are weak and were only detectable after highly sensitive earthquake monitoring networks were deployed. Unlike ordinary earthquakes that grow explosively in size with increasing duration, slow earthquakes, whether large or small, grow at a constant rate proportional to their duration. This raises the interesting question: What puts the brake on slow earthquakes? There are many other important unanswered questions about slow earthquakes, but the one most relevant for this discussion is their possible relation to ordinary earthquakes.

Slow earthquakes occur on the deep extension of large faults (Figure 4.7). This location is "strategic" for earthquake prediction because the adjoining, shallower

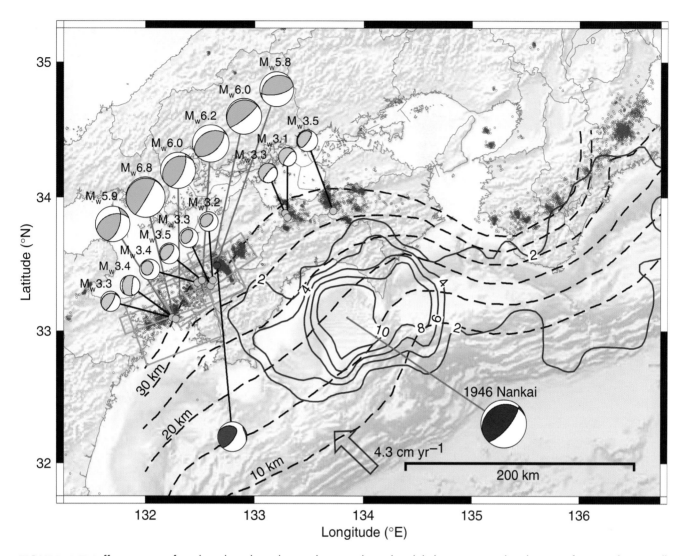

FIGURE 4.7 Different types of earthquakes along the Nankai Trough, under Shikoku, Japan. Red and orange features show small low-frequency earthquakes (M < 2) and very-low-frequency earthquakes (magnitudes shown), respectively. Green rectangles and focal mechanisms show fault-slip models of larger slow-slip events (magnitudes shown). Purple features show the mechanism and slip of the M 8 1946 Nankai earthquake. The top of the Philippine Sea plate is shown by dashed contours. The blue arrow represents the direction of relative plate motion in this area. The slow earthquakes occur on the down-dip extension of the fault that ruptured in the 1946 earthquake. SOURCE: Ide et al. (2007). Reprinted by permission from Macmillan Publishers, Ltd.: *Nature*, copyright 2007.

parts of these faults generate the dangerous earthquakes we are more familiar with. Because of their location and sense of slip, slow earthquakes ought to drive the dangerous part of the fault toward failure. At least in theory, slow earthquakes have the potential to trigger large earthquakes. For this reason alone they merit intense study. Their recent discovery also demonstrates that there is still much to be learned about earthquakes and that further fundamental discoveries are sure to lie in our future.

Tsunamis

Tsunamis are generated by shallow subduction zone earthquakes and large submarine landslides (Satake, 2007). While extremely fast compared to wind-driven ocean waves, tsunami waves are more than 10 times slower than seismic waves. This is a big advantage for early warning systems, which have been operational for decades. Tsunami warnings rely on rapid analysis of seismic waves and sea-level information from tide gauges and ocean buoys. A Pacific-wide tsunami warning system was established after the 1946 Aleutian earthquake and became an international effort after the 1960 tsunami. The 2004 Indian Ocean tsunami, which killed over a quarter of a million people, initiated efforts to build similar warning systems in the Caribbean and the Indian and Atlantic oceans.

Early warning is more difficult for tsunamis that are generated by local earthquakes. In the 1983 Japan Sea earthquake, the Japan Meteorological Agency issued a warning only 12 minutes after the earthquake, yet that was 5 minutes after the tsunami struck. Japan's warning system is now capable of issuing warnings within 3 minutes of a large earthquake.

Although warning systems can provide accurate estimates of when a tsunami will arrive, predictions of wave amplitudes and coastal run-up are less precise. Progress is limited by an incomplete understanding of the hydrodynamics of tsunami propagation and run-up, as well as uncertainties in tsunami excitation. For example, subduction zone earthquakes as large as the 1946 Aleutian event occur frequently yet only rarely produce big tsunamis.

Geological studies can constrain the nature and frequency of past tsunamis, which is key information for making long-term forecasts of future events. For example, tsunami-deposited sand in northern Japan indicates that prehistorical tsunamis propagated as far as 3 km inland—2 km farther than recent tsunamis—and occur on average every 500 years. Bathymetric data collected near the Hawaiian islands suggest that volcanic collapse events over the past 4 million years have generated some of the largest known landslides (Moore et al., 1994), resulting in waves as high as 300 m above sea level. A similar event today has the potential to trigger tsunamis that would be catastrophic on a scale unprecedented in human history.

Volcanic Hazards

Volcanism poses hazards both from the direct effects of eruptions—lava flows, hot ash flows, heavy ash fall—and from secondary effects, such as tsunamis, landslides, and hot mudflows. Direct effects may be catastrophic, as demonstrated by the 1902 eruption of Mont Pelée, in Martinique, whose eruptive blast shocked the world with its rapidity and destructive extent (e.g., Scarth, 2002). Examples of severe secondary effects include the pyroclastic flows and tsunamis that killed more than 36,000 people after the 1883 eruption of Krakatau volcano, in Indonesia (Simkin and Fiske, 1983), and more recently, the mudflows produced by the 1985 eruption of Nevado del Ruiz volcano in Colombia that killed 23,000 people (Voight, 1990).

Although volcanoes have drawn the interest of naturalists since ancient times (Pliny the Elder died in AD 79 while observing a violent eruption of Mount Vesuvius), the first research aimed at predicting volcanic eruptions began with the establishment of the Vesuvius Observatory in Italy in the mid-1800s and the Hawaiian Volcano Observatory soon after the 1902 eruption in Martinique. Volcano monitoring and prediction accelerated after the 1980 eruption of Mount St. Helens, in Washington state (Figure 4.8), which focused global attention on the inherent difficulties of forecasting explosive volcanic activity. It also provided a natural laboratory for intensive study that spawned conceptual advances about volcanic eruption mechanisms and effects and provided a model for multidisciplinary approaches to volcanology. Since then other major eruptions have provided more insights, instruments for monitoring volcanoes have evolved rapidly, and a growing number of volcanoes are being monitored in real time. This abundant monitoring information has improved our understanding of the processes that move magma and associated gases from deep within the crust to the surface. It has also led to a few successful predictions of volcanic eruptions (e.g., 1991 eruption of Mount Pinatubo), and there are a growing number of cases where volcano observatories have been able to provide eruption warnings (e.g., Mount St. Helens) that have proven useful for protecting life and property. Although we have substantially more capability to predict volcanic eruptions than earthquakes (or at least to provide useful warnings of possible eruption), we do not yet understand many aspects of how volcanoes work, and we cannot predict reliably exactly when a volcano will erupt, how large or violent the eruption will be, or how large an area around the volcano will be affected.

FIGURE 4.8 Photographs showing classic volcanic eruption styles of two U.S. volcanoes. (Left) Fire fountain activity at Kilauea volcano, in Hawaii, on September 19, 1984. SOURCE: U.S. Geological Survey photograph by C. Heliker. (Right) Cataclysmic explosion of Mount St. Helens volcano, in Washington, on May 18, 1980. SOURCE: U.S. Geological Survey photo by Austin Post.

What Controls Eruption Size, Frequency, and Style?

The plate tectonics revolution of the 1960s profoundly changed volcanology, along with Earth science in general (Question 5), by providing a paradigm to explain the locations of volcanoes, their composition, and, by extrapolation, their eruptive style. Most of our planet's volcanism occurs beneath oceans, where basaltic magma is generated along the extensive network of midocean ridges (Question 4). Basaltic lava flows also typify volcanic activity at oceanic islands, such as the Hawaiian chain, which are located above deep mantle plumes. Less common, although more visible and hazardous to human populations, are eruptions of volcanoes that overlie subduction zones. Here the downgoing lithosphere provides water and other volatiles that combine with the hot surrounding mantle to generate hydrous melts. The resulting magma rises through the upper mantle along linear belts above subduction zones to pond at the mantle-crust boundary or the deep crust and then cools and partially crystallizes to form less dense, higher-SiO_2 magma that is buoyant enough to rise further in the crust. This low-density magma may either erupt or accumulate within the crust, typically 5 to 10 km below the surface, as a large magma reservoir that remains liquid for tens of thousands of years as it slowly cools and solidifies. The evolution of magma is affected strongly by how much water, CO_2, and other gases it contains. Magma produced in subduction zones typically holds 10 to 100 times more of these volatile components than magma formed at midocean ridges or in mantle plumes.

Over the past two decades our understanding of the origin of magma has greatly expanded (see also Question 4), but one of the greatest difficulties in predicting volcanic eruptions is that magma has many possible fates. If only a small amount is produced, it may just cool and refreeze at depth, leaving little evidence of a magma-forming event. However, if enough liquid is formed, it will probably rise toward the surface, being less dense than the solid rock around it. In some volcanic systems, like the Hawaiian volcanoes, the lava pours onto the surface at about the same rate it is formed deep inside Earth. In the ongoing eruption of Kilauea, lava has spewed out at a rate of about 0.1 km^3/yr for the past 24 years (Heliker and Mattox, 2003). Magma is thought to be produced within the Hawaiian mantle plume at about 0.2 km^3/yr, but some erupts through the Mauna Loa volcano and the submarine Loihi volcano, which are also active. The eruption frequency of subduction zone volcanoes is highly variable. Most erupt only once every hundred years, or even less frequently, even though magma is probably produced continuously at depth, while some erupt much more frequently. The rate of magma production in subduction zones is likely much lower than in Hawaii (estimates are closer to 0.001 to 0.01 km^3 per year per volcano; Davies and Bickle, 1991; Davidson and DeSilva, 2000), which must affect how frequently eruptions can occur. However, we still do not know why eruptions recur when they do or why eruption intervals are not consistent even for a single volcano. And we still have only crude estimates of the ratio of intruded (not erupted) magma to erupted magma. More puzzling still is that in some volcanic systems magma can remain in crustal reservoirs for tens or even hundreds of thousands of years and then be released in catastrophic eruptions; hundreds to thousands of cubic kilometers of ash and lava may erupt within hours to weeks. These "super eruptions," such as those that have happened at Yellowstone and elsewhere in the western United States over the past 2 million years, are so enormous they can shift global climate (Rampino and Self, 1992; Jones et al., 2005). We are only now developing theories for why and how magma can be stored as liquid at shallow depth for so long and what prompts the sudden eruption of virtually all of it after extended storage (e.g., Jellinek and DePaolo, 2003).

The final stage of magma ascent toward the surface determines the nature of the resulting eruption. Slow ascent of hydrous (water-rich) melts allows time for water bubble formation, gas escape, and crystallization, and resultant eruptions are relatively quiescent and typically in the form of lava domes and flows. In contrast, rapid magma ascent and decompression causes the sudden and rapid formation and expansion of gas bubbles, which produce large explosive eruptions. However, there are no firm rules about either case. The eruptive style may change abruptly within a single eruptive episode, and even lava domes may pressurize, collapse, and/or explode without warning. This inconsistency has held back efforts to accurately predict the detailed form and timing of eruptions.

The mechanisms that trigger volcanic eruptions are not well understood. The search for trigger mechanisms has focused largely on earthquakes and tides. However, some volcanic eruptions might be controlled by effects as subtle as weather. For example, most eruptions of the Pavlof Volcano in Alaska occur in the autumn and winter months (McNutt, 1999). This correlation suggests that low-pressure weather systems and storm winds raise the water level around the volcano, increasing compressive strain and effectively squeezing out magma like toothpaste out of a tube. Such seasonal fluctuations account for 18 percent of the historical average monthly eruption rate of volcanoes around the Pacific ring of fire (Mason et al., 2004).

Field mapping techniques have enabled geologists to relate volcanic deposits to the processes that formed them. In the past few decades, highly accurate techniques have become available to take the pulse of volcanoes in real time. The faint, long-period, low-frequency signals generated by the flow of magma and pressurized gases through the crust can be detected by seismometers. Surface deformation caused by magma intrusion can be detected by Global Positioning System (GPS) measurements, topographic maps of growing lava domes can be constructed with centimeter-scale accuracy with aerial LIDAR measurements, and large areas can be surveyed for deformation signals with satellite-based interferometry (Box 4.1). Ground-, air-, and satellite-based measurements can track the flux of volcanic gases, which, in turn, reveal the depth and efficiency of magma degassing. We are learning to combine such data with microanalysis of volcanic material (lava and pyroclasts) to build comprehensive models of gas loss and crystallization within magmatic conduits. Even so, our understanding of magma and gas migration in the subsurface remains insufficient to accurately assess the eruptive potential of a volcano that

BOX 4.1 Monitoring Volcanoes

The primary goal of volcano monitoring is to track the movement of magma beneath a volcano and thereby predict when, and how violently, it will erupt at the surface. Three common signals are used to monitor magma movement:

1. *Occurrence of earthquakes, which are generated when magma (and/or associated gases) migration causes rocks to break.* These types of earthquakes are common in the weeks to months before a volcanic eruption. Earthquakes triggered by the sudden release of pressurized gases are typically shallow and are common in the days to hours before an eruption.

2. *Swelling of the surface of the volcano, which is caused by rising magma.* The deformation can be detected by ground-based surveying techniques, GPS stations installed around a volcano, and aerial and satellite-based surveys.

3. *Release of volcanic gases through fractures during or before magma ascent.* These gases can be measured within active fumaroles (gas vents) or spectroscopically. Both the absolute abundance and the ratio of different gas species provide information on the location of gas release and the extent to which the magmatic system might be accumulating excess gases.

One of the most promising new techniques for improving our understanding of magma transfer into the upper crust is interferometric synthetic aperture radar (InSAR). By combining images taken months to years apart, we can detect subtle changes in elevation. For example, images of South Sister volcano, in Oregon, taken in August 1996 and October 2000 show that an area to the west of the volcano inflated by about 10 cm. This observation stimulated closer monitoring of the area by other methods, and it has been documented that the uplift has since continued more slowly, at about 2.5 cm/year, and most likely indicates intrusion of magma about 7 km below the surface. Because the intrusion has been accompanied by very little seismicity, it would not have been discovered without InSAR. Although this intrusion is unlikely to produce a volcanic eruption in the near future, documentation of both the temporal and spatial patterns of magma intrusion over the next decades will greatly improve our knowledge of crustal processes related to magma transfer and storage. InSAR has proven extremely useful for detecting subtle changes in volcanoes but has limitations as a monitoring tool in active regions because of infrequent data acquisitions.

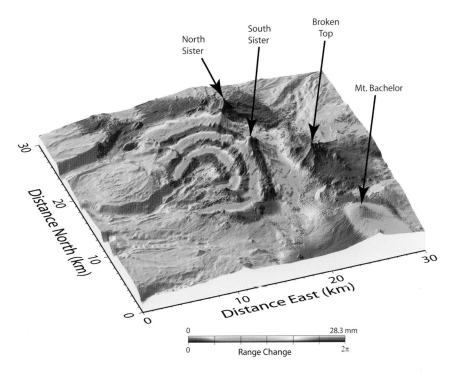

Interferogram showing uplift west of the South Sister volcano. Each full-color cycle represents 2.83 cm of range change between the ground and the satellite. SOURCE: Charles Wicks, U.S. Geological Survey.

SOURCE: <http://vulcan.wr.usgs.gov/Volcanoes/Sisters/WestUplift/framework.html>.

is showing signs of activity. Much might also be learned from synthesis of the large and growing volume of data that are available on entire volcanic arcs stretching over thousands of kilometers. Patterns in the timing and volume of eruptions at this large scale have received relatively little attention but could prove important for relating volcanic activity to rates of large-scale tectonic processes that can also be monitored with geodetic and seismic techniques.

What Aspects of Volcanic Eruptions Can Be Predicted?

If human populations are to live close to active volcanoes with a reasonable degree of safety, geoscientists must be able to (1) assess the risk of eruptive activity based on past history and (2) provide reliable predictions of eruptive potential during times of volcanic restlessness based on eruption precursors. The geological record provides information on the recurrence rates and magnitudes of large volcanic eruptions. More detailed eruptive histories of specific volcanoes have enabled some long-term predictions to be made. For example, the analysis of the history of Mount St. Helens (Crandell and Mullineaux, 1978) was notable for its accurate forecast of another eruption before the end of the century. However, it is not clear whether data from infrequent but much larger eruptions in the past can be compared with data from brief, small, recent eruptions. We know that the magnitude and destructiveness of past volcanoes have greatly exceeded anything in human history (Figure 4.9). For example, supereruptions have produced more than 2,000 km³ of pumice and ash as recently as 75,000 years ago in Indonesia (the Toba eruption; Rose and Chesner, 1987) and 2 million years ago at Yellowstone in the United States (Christiansen, 1984). This is 50 times the amount of material erupted by Tambora volcano, in Indonesia, in 1815, an eruption that caused the deaths of more than 90,000 people and disrupted global climate (Stothers, 1984; Sigurdsson and Carey, 1989). Similarly, the enormous outpourings of basaltic lava (more than 2,000 km³ during single events) in Washington and Oregon about 16 million years ago (Hooper, 1997) dwarf that of the Laki eruption of 1783 in Iceland (about 15 km³), an eruption that killed more than 9,000 Icelanders directly and unknown

FIGURE 4.9 Aerial photo of Crater Lake, which occupies the circular depression formed by the catastrophic eruption of Mount Mazama volcano about 7,700 years ago. In this eruption, about 50 km³ of ash and lava were released, about 10 times more than in the Pinatubo eruption of 1991 and about 40 times more than in the Mount St. Helens eruption of 1980. The largest eruptions documented in the geological record were 10 to 100 times larger than the Mount Mazama eruption. SOURCE: U.S. Geological Survey, <http://vulcan.wr.usgs.gov/Volcanoes/CraterLake/Locale/framework.html>.

numbers of Europeans because of crop failures and starvation (Thordarson and Self, 2003).

The largest explosive eruption monitored by modern techniques was that of Mount Pinatubo in 1991, when the release of only about 5 km³ of magma caused the collapse of the volcano's summit. Emergency evacuation of surrounding cities and towns before the event was a success for volcanic eruption prediction (Newhall and Punongbayan, 1996). Yet much larger volcanic systems such as Yellowstone have shown signs of restlessness that could, at some point, portend an impending eruption. We do not know whether we can accurately scale up modern instrumental data for Pinatubo-sized eruptions to anticipate events that may be 100 or 1,000 times larger.

Another challenge in predicting volcanic eruptions will be to combine diverse observational data sets (e.g., seismic, geodetic, infrasound, thermal, gas measurements, visual observation via webcams) to track, in real time, not only the movement of magma toward the surface but also changes in the material properties of the magma that affect its explosive potential.

This approach is currently being applied in a few cases, such as at Stromboli in Italy and Augustine in Alaska. Because every volcano is slightly different, we need predictive methodologies that apply not only to a specific volcano but to volcanoes generally. An urgent need is more widely deployed methods to monitor deep processes that may ultimately control eruptive activity (see Question 4). Such monitoring, using geodetic data primarily, has so far been applied at several volcanoes (e.g., Usu, Iwate, Miyakejima, Iwo Jima, Rabaul, Okmok, Westdahl, Akutan, South Sister, Etna, various Andean volcanoes), but there have been few chances to tie the observations to subsequent eruptions. This need applies especially to Mount St. Helens and Soufriere Hills volcano, in Montserrat, which have both been active for decades and are likely to erupt again relatively soon. Their eruptive activity requires successive inputs of magma from lower or midcrustal levels, a process that is still difficult to detect.

Summary

Thanks largely to better understanding of causes and sensitive new instrumentation, geologists have moved in recent decades toward predictive capabilities for volcanoes and, to a lesser extent, earthquakes. And yet the complexity of still-open theoretical questions and the growing human populations in threatened regions have both complicated their work and heightened its urgency.

Earth scientists have learned a great deal about predicting earthquake behavior. Plate tectonics provides a framework for understanding where most earthquakes occur and also constrains the long-term slip rate over complex fault systems. To predict the timing of individual earthquakes, however, we need to develop a deeper understanding of the factors that control the initiation and termination of fault rupture. New observational capabilities in seismology, geodesy, and geology continue to provide new insight into earthquake behavior, and new discoveries in the science of earthquakes continue apace. For ground motion prediction, high-performance computation holds forth the prospect of making physics-based simulations of earthquake strong ground motion. For all forms of earthquake prediction, it is important to find ways to validate new techniques as they are developed.

Studies of volcanic activity have also been propelled by technological developments, especially real-time seismic, electromagnetic, and geodetic probes of active subsurface processes. Improved understanding will require integrating these geophysical observations with field studies of volcanic structures and laboratory studies of volcanic materials. The ultimate objective is to develop a clear picture of magma movement: from its sources in the upper mantle to Earth's crust, where it is temporarily stored, and ultimately to the surface where it erupts. Sensitive new geophysical and geochemical techniques are improving our ability to track magma movement, and field studies of uplifted, eroded magma reservoirs and feeder systems are providing clues about how to interpret this information. Improving the safety of growing populations in volcano-prone regions will require an increase in our fundamental understanding of volcanic eruptions and public education and better planning to decrease human vulnerability to volcanic eruptions.

QUESTION 10: HOW DO FLUID FLOW AND TRANSPORT AFFECT THE HUMAN ENVIRONMENT?

Geological science has traditionally been closely tied to the assessment and discovery of natural resources such as minerals, petroleum, natural gas, geothermal water, and groundwater. More recently, geology has played a major part in understanding the fate of waste compounds and other materials released to the environment. In the future some of these waste products and byproducts, like carbon dioxide and radioactive elements from nuclear power plants, may be sequestered intentionally in geological formations. Geology is also concerned with the development of landscapes by erosion and tectonics, and increasingly this interest is focused on assessing the impacts of human activities on both the physical character of rivers and their drainage basins and the relationships between these physical characteristics, the risks of floods and landslides, and the health of ecosystems. Both of these categories of societal concern—resources and environmental impacts—are likely to increase in urgency in the future, and hence there is a continuing effort to improve access to underground resources, to maintain or manage existing resources both below ground and above ground, and to minimize or mitigate the undesirable consequences of human activities.

Perhaps the most fundamental underlying scientific theme for resource and environmental issues is the behavior of fluids in the soils, sediments, and rocks that constitute Earth's uppermost crust. Water is the most common fluid of concern. Water in the ground generally comes from water at the surface, and the behavior of surface water and ultimately, precipitation, is an important aspect of environmental geology. In addition to water, various gases, organic liquids, and both gaseous and liquid carbon dioxide are important geological fluids. Mixtures of fluids—immiscible liquids like water and hydrocarbons, gas-liquid mixtures (two-phase fluids), and mixtures of a gas phase plus two immiscible liquids (multiphase fluids)—can be particularly challenging materials to understand in natural underground settings. Some of the scientific issues associated with fluids in shallow crustal environments also apply to deeper-Earth processes, and many of them also overlap with issues of earthquake prediction, climate prediction, the evolution of continents, the behavior of volcanoes, the formation of ore deposits, and the properties of Earth materials.

Since water, as the best example, is a commodity of critical importance to humankind, and also an agent for so many important geological, chemical, physical, and biological processes, there is a continuing desire to better understand how it works—especially underground where we cannot see it directly, but also as an agent of erosion and sediment transport at the surface. Ultimately, it is desirable to be able to manipulate water and other fluids in the environment. Such manipulation has been done for millennia in the case of surface water and is also done in the subsurface, although still with modest efficiency, in petroleum extraction and subsurface remediation of contaminants. To improve our ability to control, or at least predict, the effects of subsurface fluids, and to better manage surface water and sediment, will require major advances in our understanding of how fluids transport materials and modify their environment by chemical and physical interactions.

How Do Fluids Flow in Geological Media?

The flow of fluids through soils and rocks is easily understood in the abstract but continues to present road-blocks to understanding in natural settings. We have a general understanding of how fluid moves through a granular solid (i.e., the mineral grains or rock fragments are packed together but separated by pore space), based on models of fluid flow through a medium of homogeneous grain size and pore structure. Natural materials are not homogeneous, however, especially on the 100- to 100,000-m scale of groundwater systems, but even on scales of microns to meters. The rate of flow through porous materials varies exponentially with porosity and grain size, so predicting the spatial pattern of fluid flow even in a relatively simple, but heterogeneous, porous material can be difficult. At the pore scale of individual mineral grains, surface tension also affects flow; the liquid phase present at the boundaries of multiple grains has different properties than a bulk liquid and can effectively be held in place by capillary forces. At larger scales, Earth's subsurface is composed of a variety of rock types, with greatly varying porosity and permeability, that are further complicated by faults and fractures.

When a rock medium is not granular but crystalline, the pore space is typically not visible to the naked eye and its distribution within the rock is exceedingly variable. Most of the pore space in crystalline rock is attributable to fractures, so the flow of fluid can be almost entirely limited to a few fractures that happen to be connected. Many geological media, especially volcanic rocks, are both porous and fractured. In these cases much of the fluid flow may be confined to fractures, but there is also chemical and heat exchange by diffusion (and slow flow) between the fractures and the porous rock between the fractures.

Given this battery of uncertainties, geologists have developed a number of strategies to predict fluid flow patterns in rocks, including some that are largely empirical. A more promising approach is to treat the structural variability with statistical methods, based on observations of analogous rocks that can be studied at the surface. But the flow of fluids through rocks underground remains exceptionally difficult to predict. Generally the best results are mere estimates, and even these are obtainable only from direct observations, usually by drilling into the subsurface and making measurements of returned fluids and rock cores. Still, there is cause for optimism because increasingly powerful measuring tools are being developed—using approaches such as isotope geochemistry and geophysics—and more effective mathematical modeling allows geologists to wring more information from the data obtained.

How Do Fluid-Rock Chemical and Biological Interactions Affect Fluid Flow?

As fluids flow through soils and rocks, chemical reactions inevitably occur with the minerals of the rocks, sometimes catalyzed by microorganisms. The most familiar interaction is adsorption, or ion exchange, by which ions carried in solution in water are adsorbed and desorbed from mineral surfaces. This process, which happens everywhere in nature, has been successfully exploited by humans to create water purification systems. Fluids moving through rocks also act as weak acid solutions, often due to dissolved carbon dioxide, that slowly dissolve the original minerals, which are then replaced by secondary minerals such as rusty iron oxides and clays. As rocks and soils chemically react with fluids, changes occur not only in mineralogical and chemical composition, but also in ion exchange and hydrological properties. For soils, the activities of plants, animals, and microbes are important. In deeper groundwater systems, where temperatures are higher and fluids can be more corrosive, chemical reactions can be quite fast. But because chemical reactions between fluids and minerals occur only at mineral surfaces, the structure of the fluid flow through rocks and the geochemistry are inextricably linked. If fluid flow is confined to a few fractures, it may be fast, with little contact area between fluid and minerals and little chemical interaction. If there is grain-scale porous flow, however, flow velocity will be low, the contact area large, and fluid-rock interaction extensive.

Geological studies of fluid flow, chemical reactions, and their interplay are grouped under the heading of reactive chemical transport (e.g., Steefel et al., 2005; Figure 4.10). A major goal of this subfield is to describe, with advanced computational techniques, how the characteristics of fluid-rock systems affect their physical, chemical, and biological development. The computer models require large inputs of basic materials property data, and the complexity of the interactions is a conceptual as well as a computing challenge. One crucial feature of the models is mineral surface properties and their role in chemical reaction kinetics, which are increasingly explored at synchrotron X-ray facilities.

Other inputs come from benchtop experiments that produce and observe coupled processes in a realistic, controlled environment. In addition to modeling, efforts are being made to document the role of microbes in altering mineral surfaces and chemical microenvironments (Figure 4.10). And the role of hydrology in chemical reactions is being approached with a combination of numerical models, such as approaches that include multiphase flow in complex geometries and microfluidic experiments, both of which can address the roles of chemical transport and pore structure on chemical reactions. For multiphase fluids there are additional considerations because the presence of each phase interferes with the flow of the other phases, and the detailed distribution of each phase within the pores can affect the surface area that is available for fluid-rock chemical interactions. There is also partitioning of chemical elements between multiple flowing phases (e.g., gas, oil, water), which is important in many subsurface processes but difficult to model because of its dependence on the physical relationships between the phases.

How Do Thermal and Mechanical Reactions Affect Fluid Flow?

Chemical reactions are not the only processes that complicate fluid flow. As fluids move through rocks they redistribute heat as well as material, and both the heat and materials affect the subsequent fluid flow. For example, buoyant upwelling of groundwater heated by magma can cause rainwater that has percolated into the ground to circulate to depths of several kilometers in areas of active volcanism and mountain building, as well as in sedimentary basins. At midocean ridges, cold seawater circulates through hot rocks to depths of several kilometers, and magma at the shallow depths of midocean ridges causes such rapid heating of water that it is expelled back into the ocean at temperatures above 350°C. Base metal ore deposits associated with magmatic intrusions in the crust are products of "fossil" hydrothermal systems where circulating water attained temperatures of 200°C to over 500°C (Hedenquist and Lowenstern, 1994; Sillitoe and Hedenquist, 2003). Some of these systems persisted for tens to hundreds of thousands of years at depths of 3 to 10 km. Any magma that makes

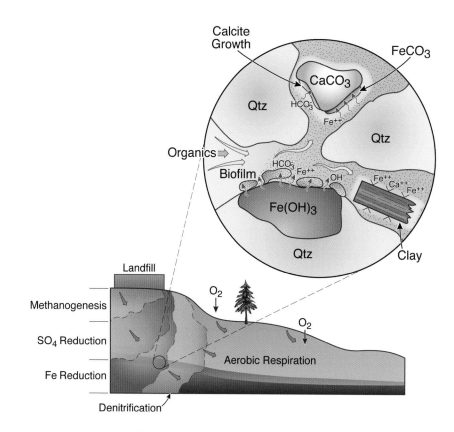

FIGURE 4.10 Schematic illustration of coupled transport, chemical, and biological processes in a hypothetical aquifer downstream of an organic-rich landfill. Closest to the landfill is a zone of methane generation, which is progressively followed downstream by sulfate reduction, iron reduction, denitrification, and aerobic respiration that develop as the flowing fluid becomes progressively oxidized by mixing with oxygenated water. Within the iron reduction zone, a pore-scale image (magnified about 10,000 times relative to the cross section) is shown in which the influx of dissolved organic molecules provides electrons for iron reduction mediated by a biofilm. Dissolution of the organic phase leads to the release of Fe^{2+}, HCO_3^-, and OH^- into the pore fluid, which then causes siderite or calcite to precipitate, reducing the porosity and permeability of the material. Sorption of Fe^{2+} may also occur on clays, displacing other cations originally present on the mineral surface. Where reactions are fast relative to local transport, gradients in concentration, and thus in reaction rates, may develop at the pore scale. SOURCE: Steefel et al. (2005). Copyright 2005 by Elsevier Science and Technology Journals. Reproduced with permission.

its way to within several kilometers of Earth's surface will stimulate groundwater convection, and it is now believed that this convection plays a major role in accelerating the cooling and crystallization of magma in the crust (Fournier, 1999).

Heat transfer can affect fluid flow in several ways. Simple heating can cause the rocks to swell and may cause fractures to close, decreasing permeability and slowing flow. Water that is heating, however, can also become pressurized as it expands, which can fracture the rocks or expand and lengthen existing cracks, thereby increasing permeability and flow. Alternatively, thermal contraction of rocks due to cooling by infiltrating groundwater will also induce fracturing and promote permeability increases (Majer et al., 2007). Water that

is warming is usually dissolving minerals; water that is cooling tends to precipitate minerals. Dissolution and precipitation both affect permeability and compete with temperature changes and hydrofracture in modifying fluid flow (Haneberg et al., 1999). If hot water is close to Earth's surface and hence at low confining pressure, it can boil, and this phase change introduces additional complications. Water vapor cannot hold as much dissolved rock as hot water, so boiling tends to cause mineral precipitation. Boiling also lowers a fluid's viscosity and leads to a two-phase fluid. Some geothermal systems are hot and deep enough to support supercritical fluids, whose properties and behavior are much less well known than those of their subcritical counterparts. Better knowledge of these obscure regimes may be essential for understanding geothermal systems (Fridleifsson and Elders, 2004).

Geothermal fluid-rock circulation systems typically have scales of meters to tens of kilometers, which ensures that they will encounter a range of temperatures, pressures, rock compositions, and permeability. In real systems the fluids are often saline, acidic, or toxic, and high temperature and pressure gradients result in rapid mineral precipitation that reduces porosity and fluid flow. Nevertheless, it is highly desirable to understand and predict their behavior for a variety of practical and scientific reasons. For example, there is an abundance of hot rock not far below the surface in the western United States (Blackwell and Richards, 2004), and if water, CO_2, or some other fluid could be circulated through it and returned to the surface, a sizable amount of thermal energy could be harvested. So far, efforts to do this have been only partially successful because the evolution of this thermal-hydrological-mechanical-chemical (THMC) system is highly complex, and we lack the expertise to control these processes in a way that permits manipulation of fluid flow in the subsurface (MIT, 2006).

An interesting example of a proposed man-made THMC system, which exhibits many of the complexities of fluid-rock systems in a compact form, is the planned underground nuclear waste repository at Yucca Mountain, Nevada. The radioactive materials that would be stored in the ground produce heat, and both models and experiments show that this heat will generate groundwater convection, boiling, mineral dissolution and precipitation, as well as potentially corrosive conditions around the waste canisters themselves (see Box 4.2). Each of these effects is understood individually at a reasonable level, but the evolution of the overall system is sufficiently uncertain that it affects our assessment of the risk of burying the waste. Other examples of THMC systems are those used to enhance petroleum recovery, where steam or other fluids are pumped into the ground to produce lower-viscosity oil, to enhance permeability, and to push residual oil toward existing wells.

Can the Behavior of Subsurface Fluids Be Predicted?

Because of the complexity of fluid-rock systems, we cannot yet predict how they will change over time—a criti-cal requirement for addressing groundwater recharge, waste movement, and other issues. This limitation means that monitoring is required, but the effectiveness of subsurface monitoring systems is still limited. One way to track subsurface fluids and processes, and still perhaps the most reliable approach, is to drill wells and take samples of fluids and rock. Drilling is expensive and time consuming, however, and can never provide a complete picture of the subsurface. However, new methods are being developed to translate the chemistry and isotopic composition of sampled fluids into physical and chemical characteristics of the regions between the well samples. For example, fluid sampling now makes possible estimates of in situ chemical weathering rates; the sources, age, and velocity of the fluids; and the importance of fracture flow and even the spacing between flowing fractures.

There is hope that noninvasive geophysical methods will yield increasing amounts of information about subsurface fluid-rock systems. Geophysical methods can help detect subsurface fluids, either from the surface or between bore holes. These methods combine electrical and mechanical signals with tomographic analysis to provide three-dimensional maps of subsurface properties. The challenge is to detect the relatively weak signals and then convert them into reliable estimates of hydrological quantities such as fluid content, fluid composition, and porosity. Figure 4.11 shows an example of tomographic imaging, which can assess the connectivity of pore spaces or determine in situ spatially distinct densities. Such imaging provides a powerful new tool for understanding the spatial characteristics of Earth materials.

Still, the uncertainties in predicting long-term fluid-rock system performance are so daunting that we need much more accurate and efficient monitoring methods. The extent to which such monitoring can be done remotely or by noninvasive methods will determine just how useful they can be in monitoring contaminated groundwater sites and other systems. In general, improvement of monitoring methods hinges on fundamental advances in the chemistry and physics of geological materials. This is because the chemical, electrical, and seismic behavior of the bulk media is often determined by the details of minerals, mineral surfaces, phase boundaries, and phase compositions. And these advances, in turn, must be optimized

BOX 4.2 Thermal-Hydrological-Mechanical-Chemical Processes in Yucca Mountain

An interesting example of coupled thermal-hydrological-mechanical-chemical (THMC) processes is the anticipated behavior of the water-under-saturated rock mass surrounding the proposed horizontal tunnels (drifts) at the Yucca Mountain site in southern Nevada. Small amounts of groundwater are present in the porous volcanic rocks, even above the water table, and the primary unknown is whether this water will enter the drifts and cause the waste canisters to corrode (Long and Ewing, 2004). Such corrosion could eventually allow the release of radioactive elements into solution, and the dissolved constituents could percolate slowly downward toward the water table and the regional groundwater aquifer.

To test the likelihood of this outcome, mathematical models have been developed to simulate the combined effects of heat and material transport around the drifts (conceptual model shown below). For these models the radioactive waste containers are considered to be only a heat source, and the amount of heat they produce can be estimated accurately. This heat will keep rock temperatures near the drifts above the boiling point of water for a considerable period of time. Boiling of groundwater would generate vapor that migrates away from the drifts and then condenses in cooler regions and drains through the fracture network. This elevated temperature and moisture redistribution would cause changes in pore water and gas compositions, as well as mineral dissolution and precipitation. Mineral dissolution and precipitation can result in porosity and permeability changes in the rock, which lead to altered flow paths and flow focusing.

The models suggest that in some circumstances very little water can enter the drift but that in other circumstances some water can enter. The scientific unknowns reflect the basic and overlapping questions that characterize the broader field of fluid flow: how to represent the processes in the model, how the rock properties change as mineral dissolution and precipitation proceed, how fast the minerals actually dissolve and precipitate, how water is distributed between fractures and the porous matrix rock separating them, and the effects of heating and cooling on fracture porosity and permeability.

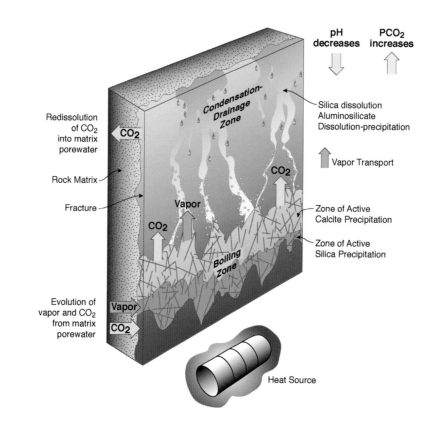

Conceptual model of processes in a fracture within volcanic tuff above a heat source with the properties of a radioactive waste container. Temperature is highest at the bottom of the fracture, nearest the heat source, and decreases upward. Heating of the fluid near the base of the fracture causes boiling (production of steam) and release of gaseous CO_2. The vapor phase rises upward due to gravity, condenses at a higher level, and flows downward as a liquid. The CO_2 dissolves into the colder groundwater near the top of the system, making it more acidic and causing it to dissolve silicates. The dissolved constituents are carried downward in the liquid phase and precipitated, eventually causing the fractures to narrow or become sealed. Although the general features of this system can be established, the time evolution is highly dependent on the rates of flow, the rates of the dissolution reactions, and resultant changes in porosity and permeability. SOURCE: Sonnenthal et al. (2001).

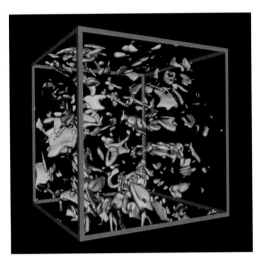

FIGURE 4.11 Tomographic image of residual fluid saturation in a sintered bead pack after free drainage. The beads have been rendered transparent and were 1.63 mm in diameter. SOURCE: Sakellariou et al. (2004). Copyright 2004 by Elsevier Science and Technology Journals. Reproduced with permission.

by improvements in data analysis and computing capabilities.

What Are the Effects of Multiple Timescales and Length Scales on Fluid-Rock Systems?

As with other Earth materials and processes, the behavior of fluid-rock systems varies enormously with length scales and timescales. Although some processes can be studied in the laboratory, experiments must generally be limited to systems that are centimeters to meters in size and days to months in duration. In this setting it is possible to characterize the average properties of fluid flow and accurately predict both flow and chemical interactions, but as we have seen throughout this report, laboratory results cannot faithfully reflect those of natural systems that are much larger and persist for thousands or millions of years. In general, larger systems exhibit faster flow, greater dispersion, and much slower chemical interactions between fluids and solids than we expect on the basis of laboratory experience.

The problems of scale are more than technicalities; they are fundamental scientific challenges, as noted also for material properties (Question 6), earthquake prediction (Question 9), and global weathering rates (Question 7). Geological features that are present at one scale—for example, faults and lithologic changes at scales of thousands of meters—are not present at smaller scales. Consequently, it is not justifiable to extrapolate material properties like hydraulic conductivity from small to large scales. With regard to fluid-mineral chemical reactions, reaction rates depend on the fluid's chemical composition, the mineral-fluid contact area, and the microscopic characteristics of the mineral surfaces (Question 6). All of these parameters can vary, and the range of variation can be large at both small and large spatial scales. Also, the chemical reactions that are significant at geological timescales proceed at ultraslow rates (Question 7), and it is obvious that the factors that control these rates are not the same as those that control laboratory reactions that proceed a million or a billion times faster.

Variation of material properties and reaction rates at different timescales and length scales is an important issue for large-scale geological sequestration of carbon dioxide (Box 4.3). Typical plans are to inject CO_2 into sedimentary rock formations deep underground at hundreds of sites over periods of tens of years. The injected CO_2, which is lighter than saline aqueous fluid, can displace the fluids but at the same time will tend to mix with the ambient fluid to produce a carbonic acid-rich dense fluid. Since the CO_2 must be retained underground for hundreds of years for geological sequestration to be effective, and because the fluids will be confined only by geological barriers, it is important to understand how the fluid will move and react with the rocks and to have the capability to monitor the movement and reactions (DOE, 2007).

Can the Effects of Water on Earth Processes Be Predicted?

Water, in both gaseous and liquid forms, is a uniquely pervasive fluid in the ways it supports life and otherwise influences the structure and evolution of the planet—and yet we only partially understand most of these processes. For example, humans depend on the balance between the extraction of groundwater and the recharge of groundwater reservoirs, but the factors affecting this balance are complex. They depend on how rainfall is partitioned between evaporation back to the atmosphere, surface runoff, and infiltration into deeper reservoirs where evaporation is no longer important. In

BOX 4.3 Geological Sequestration of CO_2

Several techniques have been proposed to capture CO_2 at the source and sequester it from the atmosphere. One approach under active investigation is storage in geological reservoirs. Current feasible options for geological sequestration include oil and gas reservoirs, coal beds, and saline formations (i.e., saline aquifers and brine-saturated sedimentary rock). Although nature has stored CO_2 in these geological structures for millions of years, the human use of this technique has other advantages. For example, injecting CO_2 into an oil or a natural gas reservoir can enhance hydrocarbon production, and about 35 million tons of CO_2 per year is already used for this purpose in the United States (Stevens et al., 2000). Limited field tests suggest that CO_2 injection would also enhance extraction of methane from coal beds by displacing methane with CO_2. Although these techniques have the potential to enhance resource recovery and offset the costs of CO_2 capture, transport, and injection, questions of reservoir availability (for oil and gas) and technological readiness (for coal beds) limit their widespread use.

CO_2 can also be injected into saline formations in sedimentary basins and on the continental shelves and trapped by displacement or compression of brine in the porous rocks. Disposal under several hundred meters of deep-sea sediment is another option. The limiting factors on storage volume include sediment layer thickness and permeability, as well as the potential for ground perturbations, such as landslides (House et al., 2006).

To what extent will CO_2 move within or beyond the geological formation? What physical and chemical changes are likely to occur in the formation when CO_2 is injected? A better understanding of the implications of CO_2 injection and sequestration is critical to determining its viability as a mitigation option for atmospheric CO_2 emissions. Key questions that must be answered include the location, capacity, and availability of storage sites; the permanence of storage; and the risks to humans and ecosystems. Long-term monitoring, measurement, and verification technologies must also be developed to improve the storage prediction models used to estimate storage capacity and to design storage areas.

Leakage of CO_2 from storage sites, as well as migration within the sites, could pose local and global environmental hazards; in 1986 the sudden venting of CO_2 from the bottom of Lake Nyos, Cameroons, killed 1,800 people. Escaped CO_2 can infiltrate the shallow subsurface, with potential adverse effects on groundwater chemistry, the vadose zone, and ambient air quality above ground. Large-scale storage failure could return CO_2 to the global atmosphere. Understanding the fate of these gases and fluids on timescales of thousands of years and longer is crucial to decisions on the wise use of carbon-based fuels.

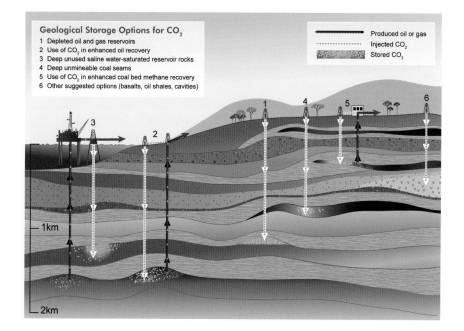

Approaches for geological sequestration of carbon dioxide. SOURCE: Cooperative Research Centre for Greenhouse Gas Technologies (CO2CRC). Used with permission.

regions of thick vegetation, plants recycle much of the water back to the atmosphere. In arid regions there is both little rainfall and little vegetation, which in some ways simplifies the analysis, but because little rainfall infiltrates in these regions, it is difficult to estimate the amount accurately. In addition, water in the form of rain is a weathering agent, which in combination with microbial processes and dust deposition produces soil. Because soil is removed or modified by land use changes, the rates of soil formation are critical in predicting the future character of the land available for agriculture, home sites, and industry. Finally, the potential effects of global warming on groundwater availability and quality may alter the living patterns of human populations in the future, but by how much is virtually unknown (IPCC, 2007b).

Some of the ways in which water influences gross planetary structure and evolution are equally obscure. We know that the presence of water in the subsurface, in particular its retention in soils and unconsolidated rocks, tends to lower internal friction and promote landslides. Water also plays a key role in determining the strength of faults and the deformability of rocks deep in the crust and mantle (Question 9), but the details are too complex to facilitate the prediction of earthquakes. Little is yet known about how fluids, primarily water and carbon dioxide, are distributed in and move through the continental crust at depths greater than a few kilometers. Water also promotes melting in planetary interiors (Question 4), and at moderate pressure and temperature inside Earth, water and silicates may be completely miscible. The behavior of such high-temperature hydrosilicate fluids is poorly known but is likely to be important for understanding both the distribution of water within planets and the origin of magmatism. Magmas constitute a class of fluids whose flow and thermal, mechanical, and reactive behavior are only crudely understood.

Beyond Earth, water and other liquids may be important for understanding geological history and the present structure of other planets and moons in the Solar System. A compelling example is Mars, where the past and present distributions of water are guiding our search for other life forms. Subsurface water is also likely to have been critical to Martian landforms, and the amount of water in the Martian mantle and deep crust is likely to have influenced the planet's evolution.

Can Landscapes Be Managed to Sustain Human Populations and Ecosystems?

Water flowing at the surface in rivers and streams transports dissolved ions, sediment, and organic material and constitutes a longstanding focus of geological study. Surface water, through erosion and sediment redistribution, is the primary sculptor of Earth's landscapes—or was until the rapid population growth over the past century. We are living on a planet that we have "engineered" over the millennia. Humans have caused massive changes in the shape of landscapes as well as the distribution of plants, animals, water, sediments, and chemicals. These changes have been caused by resource extraction, as well as attempts to ensure water availability, promote agriculture, build roads, and decrease the risk of floods and landslides. Recently we have learned that these changes generate new risks in the longer term. If we are to protect and sustain the planetary systems that provide us with essential services, we must base our future "engineering" decisions on a thorough understanding of the fundamental processes that govern how Earth works.

A sustainable landscape is one that supports the continued use of resources while maintaining critical natural processes and ecosystem functions. Humans will always need to extract resources, but minimizing damage to the environment will require a more effective capability to link specific actions to quantifiable consequences. In a single watershed, for example, actions such as timber harvesting, plowing, and road construction are known to cause downstream changes in sediment transport, water flow, and nutrient availability. But models of watershed processes, linking physical, biological, and geochemical processes, are poorly developed. New technologies are emerging that should improve those models, including high-resolution topographic data from airborne laser swath mapping, which can be used to measure even small changes in landscape morphology; new sediment tracers and dating methods; inexpensive wireless sensors that enable spatially distributed, intensive monitoring; and more powerful computational capabilities that allow the integration of these diverse data into mathematical models.

The desire to restore landscapes and ecosystems to their "predisturbance" states has led to an emerging field of restoration geomorphology. A key question is whether it is possible to help a dynamic landscape persist through human-induced changes and retain its most important and desirable attributes. A good example is stream restoration (Bernhardt et al., 2005), which presents a surprisingly complicated set of objectives. In a typical situation the desired state might be a laterally migrating, self-maintaining stream channel that passes the sediment it receives, rather than allowing it to accumulate in undesirable places; maintains habitat for plants and animals; and maintains its dissolved load and nutrient content at appropriate levels. Although we are learning how to address some of these objectives, we lack mechanistic models for river channels that represent their morphology, sediment load, and interaction with vegetation. And even a good design for current conditions might not be useful through flood-drought cycles and longer term climate changes. Another example of landscape change is dam building. While this kind of change is completely human caused and initially local, it is now recognized to have effects that are global in scale (Syvitski et al., 2003, 2005; see Box 4.4).

Given the inevitability of environmental change, whether natural or human induced, stream systems need to be managed for the desirable ecosystem characteristics even if, for example, sea level rises, precipitation changes, or mountain glaciers disappear. For example, global warming brings permafrost melting in polar regions, along with a range of hydrological, ecological, and geochemical changes (Chapin et al., 2006). Because warming will continue well into the future no matter how we attempt to manage greenhouse gases today, human societies need the capability to predict the consequences and take actions that preserve functions and resources (e.g., see Box 4.3).

Hazards from surface processes include landsliding, flooding, and coastal retreat. Hazard mitigation has traditionally relied on the use of maps that delineate some aspect of risk, but such maps tend to rely on the intuitive skill of the mapmaker and are typically based on a fixed environmental state. This means that the maps rapidly become inaccurate. With advances in weather

and climate forecasting, the availability of digital topography, and improved understanding of processes, hazard prediction is becoming spatially explicit, up to date, and much more useful for mitigation efforts. With today's 10-day forecasts of weather, flood forecasts are becoming commonplace as well, although not yet achieving good spatial extent and accuracy. Scientists are also beginning to forecast landslides in response to predicted rainfalls, but they still lack the ability to predict landslide size, location, travel distance, or speed. Sea-level rise, changes in storminess, and reductions in sediment due to dams may influence the effects of large storms on lowland river, delta, and coastal systems. However, we cannot yet predict how sea-level rise will affect levees or the flood heights on lowland rivers or determine whether artificial levees could be removed while still retaining flood protection. Answering these and many other such questions will require a body of field studies, experiments, theory, and numerical modeling sufficient to build the predictive science of watershed resiliency and hazard mitigation.

Summary

Our ability to manage natural resources, safely dispose of wastes, and sustain the environment depends on our understanding of fluids, both at the surface and below ground. In particular, we need a better grasp of how fluids flow, how they transport materials and heat, and how they interact with and modify their surroundings. The list of significant fluids begins with water, the most abundant and important Earth fluid, and includes steam, hydrocarbons, liquid and gaseous carbon dioxide, other organic liquids, and multiphase fluids (gas plus liquid, immiscible liquids, and gas plus immiscible liquids). For subsurface processes we need to understand how these fluids are distributed in heterogeneous rock and soil formations, how fast they flow, and how they are affected by chemical and thermal exchange with the host formations. At Earth's surface we are concerned with the flow of water in rivers and streams, how stream erosion and sediment transport change landscapes, and how human activities and climate change affect the evolution of streams and landscapes.

Decades of research have brought substantial knowledge about the flow and transport of fluids, but application of this knowledge is strained by increased

BOX 4.4 Global Impact of Dams

Humans have constructed more than 45,000 dams above 14 m in height, which together are capable of holding back about 15 percent of the total global annual river runoff (Vörösmarty et al., 2003). Dams have reduced the total amount of sediment carried to the ocean by about 20 to 30 percent, even though human activities have increased the total sediment production by 30 percent. And dam building continues; between 160 and 320 new dams are built annually, especially in Asia. Dams cause major changes to local, and ultimately worldwide, physical, chemical, and ecological systems and in many cases simply terminate river ecosystem functions.

Despite the negative impacts of dams, demand for power, flood control, and water supply means that many will remain and more will be built. Nonetheless, much can be done to preserve the desired physical, chemical, and ecological characteristics of affected watersheds. For example, the release of water from reservoirs could be designed to mimic important natural functions, such as sediment transport, recruitment of riparian vegetation, and fish reproduction. As we learn more about the long-term consequences of sediment depletion in downstream rivers and coastal environments, we can take action to compensate.

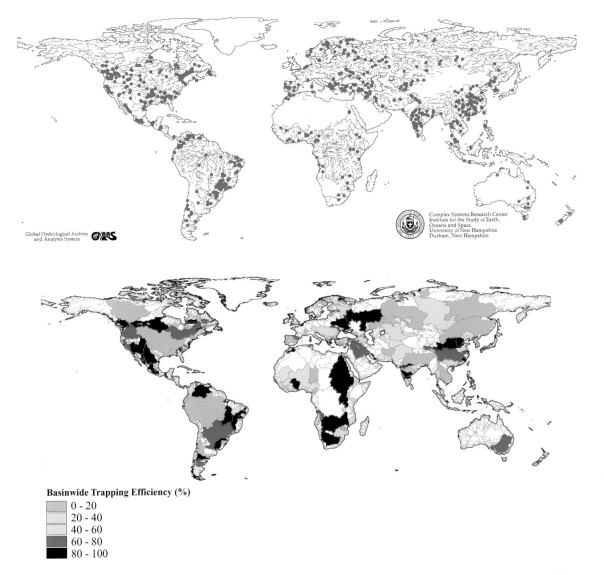

Summary of the impacts of dams on major global river systems. (Top) Geographical distribution of 633 large reservoirs (i.e., those with a storage capacity of 0.5 km³ or greater). (Bottom) Efficiency of basins in trapping suspended sediment. In some basins sediment load is severely restricted by dams along the river course; in some cases virtually all sediment is trapped. SOURCE: Vörösmarty et al. (2003). Copyright 2003 by Elsevier Science and Technology Journals. Reproduced with permission.

population, resource demands, and the environmental consequences of our own success as a species. Meeting these planetwide challenges will require a major advance in our ability to understand fluids in and on Earth, manipulate them, and monitor their whereabouts and effects. These challenges are being met by new experimental tools that can illuminate what happens at the microscopic scale on mineral surfaces, new geochemical and geophysical field techniques, and airborne and spaceborne sensors that offer an unprecedented view of how water and other fluids are shaping our planet. The ultimate objective of this research is robust mathematical models that can simulate natural fluid-bearing systems and predict far into the future how they will behave and change. Only by building and skillfully using such models will we be able to make informed decisions about the land and resources that support humankind and all life on Earth.

References

Abe, Y., 1993, Physical state of the very early Earth, *Lithos*, **30**, 223–235.

Allen, R.M., and H. Kanamori, 2003, The potential for earthquake early warning in southern California, *Science*, **300**, 786–789.

Allwood, A.C., M.R. Walter, B.S. Kamber, C.P. Marshall, and I.W. Burch, 2006, Stromatolite reef from the Early Archaean era of Australia, *Nature*, **441**, 714–718.

Alvarez, W., 1997, *T. Rex and the Crater of Doom*, Princeton University Press, Princeton, NJ, 236 pp.

Alvarez, L.W., W. Alvarez, F. Asaro, and H.V. Michel, 1980, Extraterrestrial cause for the Cretaceous-Tertiary extinction, *Science*, **208**, 1095–1108.

Andrews, D.J., 1976, Rupture velocity of plane strain shear cracks, *Journal of Geophysical Research*, **81**, 5679–5687.

Asimow, P.D., and C.H. Langmuir, 2003, The importance of water to oceanic mantle melting regimes, *Nature*, **421**, 815–820.

Atwater, B.F., 1987, Evidence for great Holocene earthquakes along the outer coast of Washington State, *Science*, **236**, 942–944.

Atwater, B.F., S. Musumi-Rokkaku, K. Satake, Y. Tsuji, K. Ueda, and D.K. Yamaguchi, 2005, *The Orphan Tsunami of 1700—Japanese Clues to a Parent Earthquake in North America*, U.S. Geological Survey Professional Paper 1707, published in association with University of Washington Press, Seattle, 133 pp.

Bada, J.L., and A. Lazcano, 2003, Prebiotic soup—Revisiting the Miller experiment, *Science*, **300**, 745–746.

Bak, P., C. Tang, and K. Wiesenfeld, 1988, Self-organized criticality, *Physical Review A*, **38**, 364–374.

Bambach, R.K., A.H. Knoll, and S.C. Wang, 2004, Origination, extinction, and mass depletions of marine diversity, *Paleobiology*, **30**, 522–542.

Banfield, J.F., and H.Z. Zhang, 2001, Nanoparticles in the environment, *Reviews in Mineralogy and Geochemistry*, **44**, 1–58.

Banfield, J.F., S.A. Welch, H. Zhang, T. Thomsen Ebert, and R.L. Penn, 2000, Aggregation-based crystal growth and microstructure development in natural iron oxyhydroxide biomineralization products, *Science*, **289**, 751–754.

Barlow, L.K., J.P. Sadler, A.E.J. Ogilvie, P.C. Buckland, T. Amorosi, J.H. Ingimundarson, P. Skidmore, A.J. Dugmore, and T.H. McGovern, 1997, Interdisciplinary investigations of the Norse western settlement in Greenland, *The Holocene*, **7**, 489–499.

Barron, E.J., and W.M. Washington, 1982, Cretaceous climate: A comparison of atmospheric simulations with the geologic record, *Palaeogeography, Palaeoclimatology, Palaeoecology*, **40**, 103–133.

Basilevsky, A.T., and J.W. Head, 2006, Impact craters on regional plains on Venus: Age relations with wrinkle ridges and implications for the geological evolution of Venus, *Journal of Geophysical Research—Planets*, **111**, E03006.

Beck, S., S. Barrientos, E. Kausel, and M. Reyes, 1998, Source characteristics of historic earthquakes along the central Chile subduction zone, *Journal of South American Earth Sciences*, **11**, 115–129.

Bekker, A., H.D. Holland, P.L. Wang, D. Rumble, H.J. Stein, J.L. Hannah, L.L. Coetzee, and N.J. Beukes, 2004, Dating the rise of atmospheric oxygen, *Nature*, **427**, 117–120.

Bennett, V.C., 2003, Compositional evolution of the mantle, in *The Mantle and Core, Treatise on Geochemistry*, **2**, R.W. Carlson, ed., Elsevier-Pergamon, Oxford, pp. 493–519.

Benton, M.J., and B.C. Emerson, 2007, How did life become so diverse? The dynamics of diversification according to the fossil record and molecular phylogenetics, *Palaeontology*, **50**, 23–40.

Bentov, S., and J. Erez, 2006, Impact of biomineralization processes on the Mg content of foraminiferal shells: A biological perspective, *Geochemistry Geophysics Geosystems*, **7**, Q01P08.

Bercovici, D., and S. Karato, 2002, Theoretical analysis of shear localization in the lithosphere, *Reviews in Mineralogy and Geochemistry*, **51**, 387–420.

Bernatowicz, T.J., S. Messenger, O. Pravdivtseva, P. Swan, and R.M. Walker, 2003, Pristine presolar silicon carbide, *Geochimica et Cosmochimica Acta*, **67**, 4679–4691.

Berner, R.A., 2006, GEOCARBSULF: A combined model for Phanerozoic atmospheric O_2 and CO_2, *Geochimica et Cosmochimica Acta*, **70**, 5653–5664.

Berner, R.A., and Z. Kothavala, 2001, GEOCARB III: A revised model of atmospheric CO_2 over Phanerozoic time, *American Journal of Science*, **301**, 182–204.

Berner, R.A., A.C. Lasaga, and R.M. Garrels, 1983, The carbonate-silicate geochemical cycle and its effect on atmospheric carbon dioxide over the past 100 million years, *American Journal of Science*, **283**, 641–683.

Berner, R.A., D.J. Beerling, R. Dudley, J.M. Robinson, and R.A. Wildman, 2003, Phanerozoic atmospheric oxygen, *Annual Review of Earth and Planetary Sciences*, **31**, 105–134.

Bernhardt, E.S., M.A. Palmer, J.D. Allan, G. Alexander, K. Barnas, S. Brooks, J. Carr, S. Clayton, C. Dahm, J. Follstad-Shah, D. Galat, S. Gloss, P. Goodwin, D. Hart, B. Hassett, R. Jenkinson, S. Katz, G.M. Kondolf, P.S. Lake, R. Lave, J.L. Meyer, T.K. O'Donnell, L. Pagano, B. Powell, and E. Sudduth, 2005, Synthesizing U.S. river restoration efforts, *Science*, **308**, 636–637.

Beroza, G., and H. Kanamori, 2007, Earthquake seismology: Comprehensive overview, in *Treatise on Geophysics*, **4**, H. Kanamori, ed., Elsevier, New York, pp. 1–58.

Bibring, J.P., Y. Langevin, J.F. Mustard, F. Poulet, R. Arvidson, A. Gendrin, B. Gondet, N. Mangold, P. Pinet, and F. Forget, 2006, Global mineralogical and aqueous Mars history derived from OMEGA/Mars express data, *Science*, **312**, 400–404.

Bizimis, M., M. Griselin, J.C. Lassiter, V.J.M. Salters, and G. Sen, 2007, Ancient recycled mantle lithosphere in the Hawaiian plume: Osmium-Hafnium isotopic evidence from peridotite mantle xenoliths, *Earth and Planetary Science Letters*, **257**, 259–273.

Blackwell, D.D., and M. Richards, 2004, *Geothermal Map of North America*, 1 sheet, scale 1:6,500,000, American Association of Petroleum Geologists, Tulsa, OK, available at <http://www.smu.edu/geothermal/heatflow/geothermal_all_us_clipped_150dpi_pagesize_legend.gif>.

Bogard, D.D., and P. Johnson, 1983, Martian gases in an Antarctic meteorite? *Science*, **221**, 651–654.

Bond, G., B. Kromer, J. Beer, R. Muscheler, M.N. Evans, W. Showers, S. Hoffmann, R. Lotti-Bond, I. Hajdas, and G. Bonani, 2001, Persistent solar influence on North Atlantic climate during the Holocene, *Science*, **294**, 2130–2136.

Boss, A.P., 2002, Formation of gas and ice giant planets, *Earth and Planetary Science Letters*, **202**, 513–523.

Boyet, M., and R.W. Carlson, 2005, ^{142}Nd evidence for early (>4.53 Ga) global differentiation of the silicate Earth, *Science*, **309**, 576–581.

Brandon, A.D., M.D. Norman, R.J. Walker, and J.W. Morgan, 1999, Os-186–Os-187 systematics of Hawaiian picrites, *Earth and Planetary Science Letters*, **174**, 25–42.

Brasier, M.D., O.R. Green, J.F. Lindsay, N. McLoughlin, A. Steele, and C. Stoakes, 2005, Critical testing of Earth's oldest putative fossil assemblage from the ~ 3.5 Ga Apex chert, Chinaman Creek, Western Australia, *Precambrian Research*, **140**, 55–102.

Brasier, M.D., N. McLoughlin, O. Green, and D. Wacey, 2006, A fresh look at the fossil evidence for early Archaean cellular life, *Philosophical Transactions of the Royal Society, London*, **B361**, 887–902.

Brocks, J.J., and A. Pearson, 2005, Building the biomarker tree of life, *Reviews in Mineralogy and Geochemistry*, **59**, 233–258.

Brocks, J.J., R. Buick, R.E. Summons, and G.A. Logan, 2003, A reconstruction of Archean biological diversity based on molecular fossils from the 2.78 to 2.45 billion-year-old Mount Bruce Supergroup, Hamersley Basin, Western Australia, *Geochimica et Cosmochimica Acta*, **67**, 4321–4335.

Brocks, J.J., G.D. Love, R.E. Summons, A.H. Knoll, G.A. Logan, and S. Bowden, 2005, Biomarker evidence for green and purple sulfur bacteria in an intensely stratified Paleoproterozoic ocean, *Nature*, **437**, 866–870.

Brune, J.N., 1996, Precarious rocks and acceleration maps for southern California, *Bulletin of the Seismological Society of America*, **86**, 43–54.

Bunge, H.P., 2005, Low plume excess temperature and high core heat flux inferred from non-adiabatic geotherms in internally heated mantle circulation models, *Physics of the Earth and Planetary Interiors*, **153**, 3–10.

Burbank, D.W., A.E. Blythe, J. Putkonen, B. Pratt-Sitaula, E. Gabet, M. Oskin, A. Barros, and T.P. Ojha, 2003, Decoupling of erosion and precipitation in the Himalayas, *Nature*, **426**, 652–655.

Butler, S.L., W.R. Peltier, and S.O. Costin, 2005, Numerical models of the Earth's thermal history: Effects of inner-core solidification and core potassium, *Physics of the Earth and Planetary Interiors*, **152**, 22–42.

Butler, R.P., J.T. Wright, G.W. Marcy, D.A. Fischer, S.S. Vogt, C.G. Tinney, H.R.A. Jones, B.D. Carter, J.A. Johnson, C. McCarthy, and A.J. Penny, 2006, Catalog of nearby exoplanets, *Astrophysical Journal*, **646**, 505–522.

Caldeira, K., and J.F. Kasting, 1992, Susceptibility of the early Earth to irreversible glaciation caused by carbon dioxide clouds, *Nature*, **359**, 226–228.

Caldeira, K., and M.R. Rampino, 1991, The mid-Cretaceous super plume, carbon dioxide, and global warming, *Geophysical Research Letters*, **18**, 987–990.

Calvert, A., E. Sandvol, D. Seber, M. Barazangi, S. Roecker, T. Mourabit, F. Vidal, G. Alguacil, and N. Jabour, 2000, Geodynamic evolution of the lithosphere and upper mantle beneath the Alboran region of the western Mediterranean: Constraints from travel time tomography, *Journal of Geophysical Research—Solid Earth*, **105**, 10,871–10,898.

Canfield, D.E., 2005, The early history of atmospheric oxygen: Homage to Robert A. Garrels, *Annual Review of Earth and Planetary Sciences*, **33**, 1–36.

Canup, R.M., 2004a, Dynamics of lunar formation, *Annual Review of Astronomy and Astrophysics*, **42**, 441–475.

Canup, R.M., 2004b, Simulations of a late lunar-forming impact, *Icarus*, **168**, 433–456.

Catling, D.C., K.J. Zahnle, and C.P. McKay, 2001, Biogenic methane, hydrogen escape, and the irreversible oxidation of early Earth, *Science*, **293**, 839–843.

Cavosie, A.J., J.W. Valley, and S.A. Wilde, 2005, Magmatic δ^{18}O in 4400–3900 Ma detrital zircons: A record of the alteration and recycling of crust in the Early Archean, *Earth and Planetary Science Letters*, **235**, 663–681.

Chambers, J.E., 2003, Planet formation, in *Meteorites, Comets, and Planets, Treatise on Geochemistry*, **1**, A.M. Davis, ed., Elsevier-Pergamon, Oxford, pp. 461–476.

Chambers, J.E., 2004, Planetary accretion in the inner Solar System, *Earth and Planetary Science Letters*, **223**, 241–252.

Chan, C.S., G. De Stasio, S.A. Welch, M. Girasole, B.H. Frazer, M.V. Nesterova, S. Fakra, and J.F. Banfield, 2004, Microbial polysaccharides template assembly of nanocrystal fibers, *Science*, **303**, 1656–1658.

Chapin, F.S., III, A.L. Lovecraft, E.S. Zavaleta, J. Nelson, M.D. Robards, G.P. Kofinas, S.F. Trainor, G.D. Peterson, H.P. Huntington, and R.L. Naylor, 2006, Policy strategies to address sustainability of Alaskan boreal forests in response to a directionally changing climate, *Proceedings of the National Academy of Sciences*, **103**, 16,637–16,643.

Chester, F.M., and J.S. Chester, 1998, Ultracataclasite structure and friction processes of the Punchbowl fault, San Andreas system, California, *Tectonophysics*, **295**, 199–221.

Chlieh, M., J.-P. Avouac, K. Sieh, D. Natawidjaja, and J. Galetzka, 2007, Heterogeneous coupling on the Sumatra megathrust constrained from geodetic and paleogeodetic measurements, *Eos, Transactions of the American Geophysical Union*, **88**, Fall Meeting Supplement, Abstract U54A-01.

Christensen, U.R., and A.W. Hofmann, 1994, Segregation of subducted oceanic crust in the convecting mantle, *Journal of Geophysical Research*, **99**, 19,867–19,884.

Christiansen, R.L., 1984, Yellowstone magmatic evolution: Its bearing on understanding large-volume explosive volcanism, in *Explosive Volcanism, Its Inception, Evolution and Hazards*, National Academy Press, Washington, D.C., pp. 84–95.

Cohen, R.E., ed., 2005, *High Performance Computing Requirements for the Computational Solid Earth Sciences*, 96 pp., available at <http://www.geo-prose.com/computational_SES.html>.

Consolmagno, G.J., and M.W. Schaefer, 1994, *Worlds Apart: A Textbook in Planetary Sciences*, Prentice-Hall, Englewood Cliffs, NJ, 323 pp.

Courtillot, V., A. Davaille, J. Besse, and J. Stock, 2003, Three distinct types of hotspots in the Earth's mantle, *Earth and Planetary Science Letters*, **205**, 295–308.

Crandell, D.R., and D.R. Mullineaux, 1978, *Potential Hazards from Future Eruptions of Mount St. Helens, Washington*, U.S. Geological Survey Bulletin 1383-C, 26 pp.

Darwin, C., 1859, *On the Origin of Species by Means of Natural Selection, or the Preservation of Favored Races in the Struggle for Life*, 1st ed., J. Murray, London.

Davaille, A., 1999, Simultaneous generation of hotspots and superswells by convection in a heterogeneous planetary mantle, *Nature*, **402**, 756–760.

Davidson, J., and S. DeSilva, 2000, Composite volcanoes, in *Encyclopedia of Volcanoes*, H. Sigurdsson, ed., Academic Press, London, pp. 663–682.

Davies, G.F., 1999, *Dynamic Earth: Plate, Plumes and Mantle Convection*, Cambridge University Press, Cambridge, 458 pp.

Davies, J.H., and M.J. Bickle, 1991, A physical model for the volume and composition of melt produced by hydrous fluxing above subduction zones, *Philosophical Transactions of the Royal Society of London*, **A335**, 355–364.

Derry, L.A., and C. France-Lanord, 1996, Neogene Himalayan weathering history and river $^{87}Sr/^{86}Sr$: Impact on the marine Sr record, *Earth and Planetary Science Letters*, **142**, 59–74.

Dieterich, J.H., 1979, Modeling of rock friction, 2. Simulation of preseismic slip, *Journal of Geophysical Research*, **84**, 2169–2175.

Dietrich, W.E., and J.T. Perron, 2006, The search for a topographic signature of life, *Nature*, **439**, 411–418.

Di Toro, G., D.L. Goldsby, and T.E. Tullis, 2004, Friction falls towards zero in quartz rock as slip velocity approaches seismic rates, *Nature*, **427**, 436–439.

DOE (Department of Energy), 2007, *Basic Research Needs for Geosciences: Facilitating 21st Century Energy Systems*, Report from the Workshop held February 21-23, 2007, Office of Basic Energy Sciences, 186 pp. plus appendixes, available at <http://www.sc.doe.gov/bes/reports/files/GEO_rpt.pdf>.

Donnadieu, Y., Y. Goddéris, G. Ramstein, A. Nédélec, and J. Meert, 2004, A 'snowball Earth' climate triggered by continental break-up through changes in runoff, *Nature*, **428**, 303–306.

Ducea, M.N., 2002, Constraints on the bulk composition and root foundering rates of continental arcs: A California arc perspective, *Journal of Geophysical Research—Solid Earth*, **107**, 2304.

Dudley, R., 2000, The evolutionary physiology of animal flight: Paleobiological and present perspectives, *Annual Review of Physiology*, **62**, 135–155.

Ekart, D.D., T.E. Cerling, I.P. Montañez, and N.J. Tabor, 1999, A 400 million year carbon isotope record of pedogenic carbonate: Implications for paleoatmospheric carbon dioxide, *American Journal of Science*, **299**, 805–827.

Ellsworth, W.L., and G.C. Beroza, 1995, Seismic evidence for an earthquake nucleation phase, *Science*, **268**, 851–855.

Ellsworth, W.L., A.G. Lindh, W.H. Prescott, and D.G. Herd, 1981, The 1906 San Francisco earthquake and the seismic cycle, in *Earthquake Prediction: An International Review*, D.W. Simpson and P.G. Richards, eds., Maurice Ewing Series, vol. 4., American Geophysical Union, Washington, D.C., pp. 126–140.

Engebretson, D.C., K.P. Kelley, H.J. Cashman, and M.R. Richards, 1992, 180 million years of subduction, *GSA Today*, **2**, 93–95.

Ernst, R.E., K.L. Buchana, and I.H. Campbell, 2005, Frontiers in large igneous province research, *Lithos*, **79**, 271–297.

Erwin, D.H., 2006, *Extinction: How Life on Earth Nearly Ended 251 Million Years Ago*, Princeton University Press, Princeton, NJ, 296 pp.

Falkowski, P.G., M.E. Katz, A.H. Knoll, A. Quigg, J.A. Raven, O. Schofield, and F.J.R. Taylor, 2004, The evolution of modern eukaryotic phytoplankton, *Science*, **305**, 354–360.

Farnetani, C.G., and H. Samuel, 2003, Lagrangian structures and stirring in the Earth's mantle, *Earth and Planetary Science Letters*, **206**, 335–348.

Farquhar, J., and B.A. Wing, 2003, Multiple sulfur isotope analyses: Applications in geochemistry and cosmochemistry, *Earth and Planetary Science Letters*, **213**, 1–13.

Farquhar, J., H. Bao, and M. Thiemens, 2000, Atmospheric influence of Earth's earliest sulfur cycle, *Science*, **289**, 756–758.

Fedorov, A.V., P.S. Dekens, M. McCarthy, A.C. Ravelo, M. Barreiro, P.B. deMenocal, R.C. Pacanowski, and S.G. Philander, 2006, The Pliocene paradox (mechanisms for a permanent El Niño), *Science*, **312**, 1485–1489.

Fedotov, S.A., 1965, Regularities of the distribution of strong earthquakes in Kamchatka, the Kurile islands, and northeast Japan, *Akademia Nauk SSSR, Institut Fiziki Zemli Trudy*, **36**, 66–93.

Fialko, Y., 2006, Interseismic strain accumulation and the earthquake potential on the southern San Andreas fault system, *Nature*, **441**, 968–971.

Field, E.H., 2007, Overview of the Working Group for the Development of Regional Earthquake Likelihood Models (RELM), *Seismological Research Letters*, **78**, 7–16.

Fischer, D.A., and J. Valenti, 2005, The planet-metallicity correlation, *Astrophysical Journal*, **622**, 1102–1117.

Fournier, R.O., 1999, Hydrothermal processes related to movement of fluid from plastic into brittle rock in the magmatic-hydrothermal environment, *Economic Geology*, **94**, 1193–1211.

Freeman, K.H., and L.A. Colarusso, 2001, Molecular and isotopic records of C-4 grassland expansion in the late Miocene, *Geochimica et Cosmochimica Acta*, **65**, 1439–1454.

Frey, H.V., 2006, Impact constraints on the age and origin of the lowlands of Mars, *Geophysical Research Letters*, **33**, L08S02.

Fridleifsson, G.O., and W. Elders, 2004, The feasibility of utilizing geothermal energy from supercritical reservoirs in Iceland, *Geothermal Resources Council Transactions*, **27**, 423–427.

Friend, C.R.L., A.P. Nutman, V.C. Bennett, and M.D. Norman, 2007, Seawater-like trace element signatures (REE + Y) of Eoarchaean chemical sedimentary rocks from southern West Greenland, and their corruption during high-grade metamorphism, *Contributions to Mineralogy and Petrology*, doi:10.1007/s00410-007-0239-z.

Furnes, H., M. de Wit, H. Staudigel, M. Rosing, and K. Muehlenbachs, 2007, A vestige of Earth's oldest ophiolite, *Science*, **315**, 1704–1707.

Gao S., R.L. Rudnick, H.L. Yuan, X.M. Liu, Y.S. Liu, W.L. Xu, W.L. Ling, J. Ayers, X.C. Wang, and Q.H. Wang, 2004, Recycling lower continental crust in the North China craton, *Nature*, **432**, 892–897.

Garnero, E.J., 2000, Heterogeneity of the lowermost mantle, *Annual Review of Earth and Planetary Sciences*, **28**, 509–537.

Gomes, R., H.F. Levison, K. Tsiganis, and A. Morbidelli, 2005, Origin of the cataclysmic Late Heavy Bombardment period of the terrestrial planets, *Nature*, **435**, 466–469.

Gonnermann, H.M., and M. Manga, 2003, Explosive volcanism may not be an inevitable consequence of magma fragmentation, *Nature*, **426**, 432–435.

Goodman, J.C., and R.T. Pierrehumbert, 2003, Glacial flow of floating marine ice in "Snowball Earth," *Journal of Geophysical Research*, **108**, C10, 3308.

Grice, K., C.Q. Cao, G.D. Love, M.E. Bottcher, R.J. Twitchett, E. Grosjean, R.E. Summons, S.C. Turgeon, W. Dunning, and Y.G. Jin, 2005, Photic zone euxinia during the Permian-Triassic superanoxic event, *Science*, **307**, 706–709.

Grootes, P.M., and M. Stuiver, 1997, Oxygen 18/16 variability in Greenland snow and ice with 10^3 to 10^5-year time resolution, *Journal of Geophysical Research*, **102**, 26,455–26,470.

Gutenberg, B., and C.F. Richter, 1954, *Seismicity of the Earth and Associated Phenomena*, 2nd ed., Princeton University Press, Princeton, NJ, 310 pp.

Halliday, A.N., 2003, The origin and earliest history of the Earth, in *Meteorites, Comets, and Planets, Treatise on Geochemistry*, **1**, A.M. Davis, ed., Elsevier-Pergamon, Oxford, pp. 509–558.

Halliday, A.N., 2006, The origin of the Earth. What's new? *Elements*, **2**, 205–210.

Halverson, G.P., P.F. Hoffman, D.P. Schrag, and A.J. Kaufman, 2002, A major perturbation of the carbon cycle before the Ghaub glaciation (Neoproterozoic) in Namibia: Prelude to Snowball Earth? *Geochemistry Geophysics Geosystems*, **3**, doi:10.1029/2001GC000244.

Hanczyc, M.M., S.M. Fujikawa, and J.W. Szostak, 2003, Experimental models of primitive cellular compartments: Encapsulation, growth, and division, *Science*, **302**, 618–621.

Haneberg, W.C., P.S. Mozley, J.C. Moore, and L.B. Goodwin, eds., 1999, *Faults and Subsurface Fluid Flow in the Shallow Crust*, Geophysical Monograph 113, American Geophysical Union, Washington, D.C., 222 pp.

Hansen, V.L., 2005, Venus's shield terrain, *Geological Society of America Bulletin*, **117**, 808–822.

Harris, N., 1995, Significance of weathering Himalayan metasedimentary rocks and leucogranites for the Sr isotope evolution of seawater during the early Miocene, *Geology*, **23**, 795–798.

Harrison, T.M., J. Blichert-Toft, W. Müller, F. Albarede, P. Holden, and S.J. Mojzsis, 2005, Heterogeneous Hadean hafnium: Evidence of continental crust at 4.4 to 4.5 Ga, *Science*, **310**, 1947–1950.

Haworth, M., S.P. Hesselbo, J.C. McElwain, S.A. Robinson, and J.W. Brunt, 2005, Mid-Cretaceous pCO_2 based on stomata of the extinct conifer *Pseudofrenelopsis* (Cheirolepidiaceae), *Geology*, **33**, 749–752.

Hazen, R.M., 2005, *Genesis: The Scientific Quest for Life's Origin*, Joseph Henry Press, Washington, D.C., 339 pp.

Hedenquist, J.W., and J.B. Lowenstern, 1994, The role of magmas in the formation of hydrothermal ore deposits, *Nature*, **370**, 519–527.

Heliker, C.C., and T.N. Mattox, 2003, The first two decades of the Pu'u 'O'o-Kupaianaha eruption: Chronology and selected bibliography, in *Pu'u 'O'o-Kupaianaha Eruption of Kilauea Volcano, Hawai'i: The First 20 Years*, C. Heliker, D.A. Swanson, and T.J. Takahashi, eds., USGS Professional Paper 1676, Denver, CO, pp. 1–28.

Hemming, S.R., 2004, Heinrich events: Massive late Pleistocene detritus layers of the North Atlantic and their global climate imprint, *Reviews of Geophysics*, **42**, RG1005.

Henderson, G.M., 2002, New oceanic proxies for paleoclimate, *Earth and Planetary Science Letters*, **203**, 1–13.

Herweijer, C., R. Seager, and E.R. Cook, 2006, North American droughts of the mid to late nineteenth century: A history, simulation and implication for Mediaeval drought, *The Holocene*, **16**, 159–171.

Hill, D.P., P.A. Reasenberg, A. Michael, W.J. Arabaz, G. Beroza, D. Brumbaugh, J.N. Brune, R. Castro, S. Davis, D. dePolo, W.L. Ellsworth, J. Gomberg, S. Harmsen, L. House, S.M. Jackson, M.J.S. Johnston, L. Jones, R. Keller, S. Malone, L. Munguia, S. Nava, J.C. Pechmann, A. Sanford, R.W. Simpson, R.B. Smith, M. Stark, M. Stickney, A. Vidal, S. Walter, V. Wong, and J. Zollweg, 1993, Seismicity remotely triggered by the magnitude 7.3 Landers, California, earthquake, *Science*, **260**, 1617–1623.

Hoffman, P.F., and S.A. Bowring, 1984, Short-lived 1.9 Ga continental margin and its destruction, Wopmay orogen, northwest Canada, *Geology*, **12**, 68–72.

Hoffman, P.F, and D.P. Schrag, 2000, Snowball Earth, *Scientific American*, **282**, 68–75.

Hoffman, P.F., A.J. Kaufman, G.P. Halverson, and D.P. Schrag, 1998, A Neoproterozoic snowball Earth, *Science*, **281**, 1342–1346.

Hofmann, A.W., 1997, Mantle geochemistry: The message from oceanic volcanism, *Nature*, **385**, 219–229.

Holland, H.D., 1984, *The Chemical Evolution of the Atmosphere and Oceans*, Princeton University Press, Princeton, NJ, 582 pp.

Holland, H.D., 2006, The oxygenation of the atmosphere and oceans, *Philosophical Transactions of the Royal Society B*, **361**, 903–915.

Holtzman, B.K., D.L. Kohlstedt, M.E. Zimmerman, F. Heidelbach, T. Hiraga, and J. Hustoft, 2003, Melt segregation and strain partitioning: Implications for seismic anisotropy and mantle flow, *Science*, **301**, 1227–1230.

Hooper, P.R., 1997, *The Columbia River Flood Basalt Province: Current Status*, American Geophysical Union Monograph 100, Washington, D.C., pp. 1–27.

House, K.Z., D.P. Schrag, C.F. Harvey, and K.S. Lackner, 2006, Permanent carbon dioxide storage in deep-sea sediments, *Proceedings of the National Academy of Sciences*, **103**, 12,291–12,295.

Huber, B.T., R.D. Norris, and K.G. MacLeod, 2002, Deep-sea paleotemperature record of extreme warmth during the Cretaceous, *Geology*, **30**, 123–126.

Hustoft, J.W., and D.L. Kohlstedt, 2006, Metal-silicate segregation in deforming dunitic rocks. *Geochemistry Geophysics Geosystems*, **7**, Q02001.

Hutton, J.C., 1788, Theory of the earth, or, An investigation of the laws observable in the composition, dissolution and restoration of land upon the globe, *Transactions of the Royal Society of Edinburgh*, **1**, 209–304, plates I and II.

Hyde, W.T., T.J. Crowley, S.K. Baum, and W.R. Peltier, 2000, Neoproterozoic 'snowball Earth' simulations with a coupled climate/ice-sheet model, *Nature*, **405**, 425–429.

Ide, S., G.C. Beroza, D.R. Shelly, and T. Uchide, 2007, A scaling law for slow earthquakes, *Nature*, **447**, 76–79.

IPCC (Intergovernmental Panel on Climate Change), 2005, *Carbon Dioxide Capture and Storage*, Working Group III of the Intergovernmental Panel on Climate Change, Metz, B., O. Davidson, H.C. deConinck, M. Loos, and L.A. Meyer, eds., Cambridge University Press, New York, 442 pp.

IPCC, 2007a, *Climate Change 2007: The Physical Science Basis, Summary for Policymakers*, Contribution of Working Group I to the Fourth Assessment Report, Geneva, Switzerland, in press, available at <http://www.ipcc.ch/SPM2feb07.pdf>.

IPCC, 2007b, *Climate Change 2007: Climate Change Impacts, Adaptation and Vulnerability. Summary for Policymakers*, Working Group II of the Intergovernmental Panel on Climate Change, draft, available at <http://www.ipcc.ch/SPM6avr07.pdf>.

Ishii, M., and J. Tromp, 1999, Normal-mode and free-air gravity constraints on lateral variations in velocity and density of Earth's mantle, *Science*, **285**, 1231–1236.

Jackson, J.B.C., and D.H. Erwin, 2006, What can we learn about ecology and evolution from the fossil record? *Trends in Ecology and Evolution*, **21**, 322–328.

Jeanloz, R., P.M. Celliers, G.W. Collins, J.H. Eggert, K.K.M. Lee, R.S. McWilliams, S. Brygoo, and P. Loubeyre, 2007, Achieving high-density states through shock-wave loading of precompressed samples, *Proceedings of the National Academy of Sciences*, **104**, 9172–9177.

Jellinek, A.M., and D.J. DePaolo, 2003, A model for the origin of large silicic magma chambers: Precursors of caldera-forming eruptions, *Bulletin of Volcanology*, **65**, 363–381.

Jenkyns, H.C., 2003, Evidence for rapid climate change in the Mesozoic–Palaeogene greenhouse world, *Philosophical Transactions of the Royal Society of London A*, **361**, 1885–1916.

Johnston, A.C., and L.R. Kanter, 1990, Earthquakes in stable continental crust, *Scientific American*, **262**, 68–75.

Jones, C.E., and H.C. Jenkyns, 2001, Seawater strontium isotopes, oceanic anoxic events, and seafloor hydrothermal activity in the Jurassic and Cretaceous, *American Journal of Science*, **301**, 112–149.

Jones, G.S., J.M. Gregory, P.A. Stott, S.F.B. Tett, and R.B. Thorpe, 2005, An AOGCM simulation of the climate response to a volcanic super-eruption, *Climate Dynamics*, **25**, 725–738.

Jordan, T.H., 1988, Structure and formation of the continental lithosphere, in *Oceanic and Continental Lithosphere; Similarities and Differences*, M.A. Menzies and K. Cox, eds., Journal of Petrology, Special Lithosphere Issue, pp. 11–37.

Kagan, Y.Y., and D.D. Jackson, 1991, Seismic gap hypothesis: Ten years after, *Journal of Geophysical Research*, **96**, 21,419–21,431.

Karato, S.-I., 1998, Seismic anisotropy in the deep mantle, boundary layers and the geometry of mantle convection, *Pure and Applied Geophysics*, **151**, 565–587.

Kasting, J.F., 1990, Bolide impacts and the oxidation-state of carbon in the Earth's early atmosphere, *Origins of Life and Evolution of the Biosphere*, **20**, 199–231.

Kasting, J.F., and D. Catling, 2003, Evolution of a habitable planet, *Annual Review of Astronomy and Astrophysics*, **41**, 429–463.

Kasting, J.F., H.D. Holland, and J.P. Pinto, 1985, Oxidant abundances in rainwater and the evolution of atmospheric oxygen, *Journal of Geophysical Research*, **90**, 10,497–10,510.

Kellogg, L.H., B.H. Hager, and R.D. van der Hilst, 1999, Compositional stratification in the deep mantle, *Science*, **283**, 1881–1884.

Kharecha, P., J.F. Kasting, and J.L. Siefert, 2005, A coupled atmosphere-ecosystem model of the early Archean Earth, *Geobiology*, **3**, 53–76.

Kirby, S.H., S. Stein, E.A. Okal, and D.C. Rubie, 1996, Metastable mantle phase transformations and deep earthquakes in subducting oceanic lithosphere, *Reviews of Geophysics*, **34**, 261–306.

Kleine, T., C. Münker, K. Mezger, and H. Palme, 2002, Rapid accretion and early core formation on asteroids and the terrestrial planets from Hf–W chronometry, *Nature*, **418**, 952–955.

Kleypas, J.A., R.A. Feely, V.J. Fabry, C. Langdon, C.L. Sabine, and L.L. Robbins, 2006, *Impacts of Ocean Acidification on Coral Reefs and Other Marine Calcifiers: A Guide for Future Research*, Report of a workshop held April 18-20, 2005, St. Petersburg, FL, National Oceanic and Atmospheric Administration Pacific Marine Environmental Laboratory, Contribution 2897, 88 pp.

Knoll, A.H., 2003, *Life on a Young Planet: The First Three Billion Years of Evolution on Earth*, Princeton University Press, Princeton, NJ, 304 pp.

Knoll, A.H., M. Carr, B. Clark, D.J. Des Marais, J.D. Farmer, W.W. Fischer, J.P. Grotzinger, S.M. McLennan, M. Malin, C. Schröder, S. Squyres, N.J. Tosca, and T. Wdowiak, 2005, An astrobiological perspective on Meridiani Planum, *Earth and Planetary Science Letters*, **240**, 179–189.

Knoll, A.H., R.K. Bambach, J. Payne, S. Pruss, and W. Fischer, 2007, A paleophysiological perspective on the end-Permian mass extinction and its aftermath, *Earth and Planetary Science Letters*, **256**, 295–313.

Kopp, R.E., J.L. Kirschvink, I.A. Hilburn, and C.Z. Nash, 2005, The Paleoproterozoic snowball Earth: A climate disaster triggered by the evolution of oxygenic photosynthesis, *Proceedings of the National Academy of Sciences*, **102**, 11,131–11,136.

Kump, L.R., M. Arthur, M. Patzkowsky, M. Gibbs, D.S. Pinkus, and P. Sheehan, 1999, A weathering hypothesis for glaciation at high atmospheric pCO$_2$ in the Late Ordovician, *Palaeogeography, Palaeoclimatology, Palaeoecology*, **152**, 173–187.

Kuypers, M.M.M., P. Blokker, E.C. Hopmans, H. Kinkel, R.D. Pancost, S. Schouten, and J.S. Sinninghe Damste, 2002, Archaeal remains dominate marine organic matter from the early Albian oceanic anoxic event 1b, *Palaeogeography, Palaeoclimatology, Palaeoecology*, **185**, 211–234.

Labrosse, S., J.-P. Poirier, and J.-L. Le Mouël, 2001, The age of the inner core, *Earth and Planetary Science Letters*, **190**, 111-123.

Lamb, S., and D. Sington, 1998, *Earth Story: The Forces That Have Shaped Our Planet*, Princeton University Press, Princeton, NJ, 256 pp.

Langmuir, C.H., E.M. Klein, and T. Plank, 1992, Petrological systematics of mid-ocean ridge basalts: Constraints on melt generation beneath ocean ridges, in *Mantle Flow and Melt Generation at Mid-ocean Ridges*, J. Phipps Morgan, D.K. Blackman, and J.M. Sinton, eds., American Geophysical Union, Washington, D.C., pp. 183–280.

Lewis, R.S., M. Tang, J.F. Wacker, E. Anders, and E. Steel, 1987, Interstellar diamonds in meteorites, *Nature*, **326**, 160–162.

Linn, A.M., 2006, Identifying grand research questions in the solid-Earth sciences, *Eos, Transactions of the American Geophysical Union*, **87**, 98.

Loewenstein, T.K., M.N. Timofeeff, S.T. Brennan, L.A. Hardie, and R.V. Demicco, 2001, Oscillations in Phanerozoic seawater chemistry: Evidence from fluid inclusions, *Science*, **294**, 1086–1088.

Long, J.C.S., and R.C. Ewing, 2004, Yucca Mountain: Earth science issues at a geologic repository for high-level nuclear waste, *Annual Review of Earth and Planetary Sciences*, **32**, 363–401.

Lovelock, J., 1979, *Gaia: A New Look at Life on Earth*, Oxford University Press, Oxford, 185 pp.

Maher, K., D.J. DePaolo, and J.C. Lin, 2004, Rates of diagenetic reactions in deep-sea sediment: In situ measurement using $^{234}U/^{238}U$ of pore fluids, *Geochimica et Cosmochimica Acta*, **68**, 4629–4648.

Majer, E.L., R. Baria, M. Stark, S. Oates, J. Bommer, B. Smith, and H. Asanuma, 2007, Induced seismicity associated with enhanced geothermal systems, *Geothermics*, **36**, 185–222.

Marshall, C.P., G.D. Love, C.E. Snape, A.C. Hill, A.C. Allwood, M.R. Walter, M.J. Van Kranendonk, S.A. Bowden, S.P. Sylva, and R.E. Summons, 2007, Structural characterization of kerogen in 3.4 Ga Archaean cherts from the Pilbara Craton, Western Australia, *Precambrian Research*, **155**, 1–23.

Marshall, H.G., J.C.G. Walker, and W.R. Kuhn, 1988, Long-term climate change and the geochemical cycle of carbon, *Journal of Geophysical Research*, **93**, 791–802.

Mason, B.G., D.M. Pyle, W.B. Dade, and T. Jupp, 2004, Seasonality of volcanic eruptions, *Journal of Geophysical Research*, **109**, B04206.

McDonough, W.F., 2007, Mapping the Earth's engine, *Science*, **317**, 1177–1178.

McDonough, W. F., and S.-S. Sun, 1995, The composition of the Earth, *Chemical Geology*, **120**, 223–253.

McKay, C.P., 2000, Thickness of tropical ice and photosynthesis on a snowball Earth, *Geophysical Research Letters*, **27**, 2153–2156.

McKay, D.S., E.K. Gibson, K.L. Thomas-Keprta, H. Vali, C.S. Romanek, S.J. Clemett, X.D.F. Chillier, C.R. Maechling, and R.N. Zare, 1996, Search for past life on Mars: Possible relic biogenic activity in Martian meteorite ALH84001, *Science*, **273**, 924–930.

McKenzie, D., 1985, The extraction of magma from the crust and mantle, *Earth and Planetary Science Letters*, **74**, 81–91.

McKenzie, D., J. Jackson, and K. Priestley, 2005, Thermal structure of oceanic and continental lithosphere, *Earth and Planetary Science Letters*, **233**, 337–349.

McNeil, D., M. Duncan, and H.F. Levison, 2005, Effects of type I migration on terrestrial planet formation, *Astronomical Journal*, **130**, 2884–2899.

McNutt, S.R., 1999, Eruptions of Pavlof Volcano, Alaska, and their possible modulation by ocean load and tectonic stresses: Re-evaluation of the hypothesis based on new data from 1984–1998, *Pure and Applied Geophysics*, **155**, 701–712.

Miller, G.H., E.M. Stolper, and T.J. Ahrens, 1991, The equation of state of a molten komatiite. 2. Application to komatiite petrogenesis and the Hadean mantle, *Journal of Geophysical Research—Solid Earth and Planets*, **96**, 11,849–11,864.

Miller, K.G., M.A. Kominz, J.V. Browning, J.D. Wright, G.S. Mountain, M.E. Katz, P.J. Sugarman, B.S. Cramer, N. Christie-Blick, and S.F. Pekar, 2005, The Phanerozoic record of global sea-level change, *Science*, **310**, 1293–1298.

Miller, S.L., 1953, A production of amino acids under possible primitive Earth conditions, *Science*, **117**, 528–529.

Miller, S.L., and G. Schlesinger, 1984, Carbon and energy yields in prebiotic syntheses using atmospheres containing CH_4, CO and CO_2, *Origins of Life and Evolution of the Biosphere*, **14**, 83–90.

MIT (Massachusetts Institute of Technology), 2006, *The Future of Geothermal Energy: Impact of Enhanced Geothermal Systems (EGS) on the United States in the 21st Century*, Idaho National Laboratory, Idaho Falls, available at <http://www1.eere.energy.gov/geothermal/future_geothermal.html>.

Mojzsis, S.J., T.M. Harrison, and R.T. Pidgeon, 2001, Oxygen-isotope evidence from ancient zircons for liquid water at the Earth's surface 4,300 Myr ago, *Nature*, **409**, 178–181.

Moore, J.G., W.R. Normark, and R.T. Holcomb, 1994, Giant Hawaiian landslides, *Annual Review of Earth and Planetary Sciences*, **22**, 119–144.

Moreira, M., K. Breddam, J. Curtice, and M.D. Kurz, 2001, Solar neon in the Icelandic mantle: New evidence for an undegassed lower mantle, *Earth and Planetary Science Letters*, **185**, 15–23.

Müller, R.D., W.R. Roest, J.-Y. Royer, L.M. Gahagan, and J.G. Sclater, 1997, Digital isochrons of the world's ocean floor, *Journal of Geophysical Research*, **102**, 3211-3214.

Murakami, M., K. Hirose, K. Kawamura, N. Sata, and Y. Ohishi, 2004, Post-perovskite phase transition in $MgSiO_3$, *Science*, **304**, 855–858.

Navrotsky, A., 2004, Energetic clues to pathways to biomineralization: Precursors, clusters, and nanoparticles, *Proceedings of the National Academy of Sciences*, **101**, 12,096–12,101.

Newhall, C.G., and R.S. Punongbayan, 1996, *Fire and Mud: Eruptions and Lahars of Mount Pinatubo, Philippines*, Philippine Institute of Volcanology and Seismology, Quezon City, and University of Washington Press, Seattle, 1126 pp.

NRC (National Research Council), 1987, *Earth Materials Research: Report of a Workshop on Physics and Chemistry of Earth Materials*, National Academy Press, Washington, D.C., 122 pp.

NRC, 2001, *Basic Research Opportunities in Earth Science*, National Academy Press, Washington, D.C., pp. 35–45.

NRC, 2003a, *Connecting Quarks with the Cosmos: Eleven Questions for the New Century*, National Academy Press, Washington, D.C., 222 pp.

NRC, 2003b, *Living on an Active Earth: Perspectives on Earthquake Science*, National Academies Press, Washington, D.C., 418 pp.

Nutman, A., 2006, Antiquity of the oceans and continents, *Elements*, **2**, 223–227.

Oganov, A.R., M.J. Gillan, and D.G. Price, 2005, Structural stability of silica at high pressures and temperatures, *Physical Review B*, **71**, 064104.

Ohmoto, H., Y. Watanabe, T. Otake, D.C. Bevacqua, D. Walizer, and A. Klarke, 2005, Evolution of the atmosphere, oceans, and biosphere on early Earth: Geological, geochemical, and biological approaches, Project Reports 2005, available at <http://nai.arc.nasa.gov/team/index.cfm?page=projectreports&year=7&teamID=24&projectID=1393>.

Orgel, L.E., 2004, Prebiotic chemistry and the origin of the RNA world, *Critical Reviews in Biochemistry and Molecular Biology*, **39**, 99–129.

Ozima, M., and F.A. Podosek, 1999, Formation age of Earth from $^{129}I/^{127}I$ and $^{244}Pu/^{238}U$ systematics and the missing Xe, *Journal of Geophysical Research*, **104**, 25,493–25,499.

Palme, H., and H.S.C. O'Neill, 2003, Cosmochemical estimates of mantle composition, in *The Mantle and Core, Treatise on Geochemistry*, **2**, R.W. Carlson, ed., Elsevier-Pergamon, Oxford, pp. 1–38.

Panero, W.R., and L.P. Stixrude, 2004, Hydrogen incorporation in stishovite at high pressure and symmetric hydrogen bonding in delta-AlOOH, *Earth and Planetary Science Letters*, **221**, 421–431.

Parman, S.W., M.D. Kurz, S.R. Hart, and T.L. Grove, 2005, Helium solubility in olivine and implications for high He-3/He-4 in ocean island basalts, *Nature*, **437**, 1140–1143.

Pasteur, L., 1922-1939, *Ouevres*, 7 volumes, Masson et Cie, Paris.

Patterson, C., 1956, Age of meteorites and the Earth, *Geochimica et Cosmochimica Acta*, **10**, 230–237.

Pavlov, A.A., and J.F. Kasting, 2002, Mass-independent fractionation of sulfur isotopes in Archean sediments: Strong evidence for an anoxic Archean atmosphere, *Astrobiology*, **2**, 27–41.

Pavlov, A.A., M.T. Hurtgen, J.F. Kasting, and M.A. Arthur, 2003, Methane-rich Proterozoic atmosphere? *Geology*, **31**, 87–90.

Penn, R.L., and J.F. Banfield, 1999, Morphology development and crystal growth in nanocrystalline aggregates under hydrothermal conditions: Insights from titania, *Geochimica et Cosmochimica Acta*, **63**, 1549–1557.

Penny, D., 2005, An interpretive review of origin of life research, *Biology and Philosophy*, **20**, 633–671.

Perry, H.K.C., J.C. Mareschal, and C. Jaupart, 2006, Variations of strength and localized deformation in cratons: The 1.9 Ga Kapuskasing uplift, Superior Province, Canada, *Earth and Planetary Science Letters*, **249**, 216–228.

Peters, S.E., 2005, Geologic constraints on the macroevolutionary history of marine animals, *Proceedings of the National Academy of Sciences*, **102**, 12,326–12,331.

Phillips, B.L., W.H. Casey, and M. Karlsson, 2000, Bonding and reactivity at oxide mineral surfaces from model aqueous complexes, *Nature*, **404**, 379–382.

Pierrehumbert, R.T., 2004, High levels of atmospheric carbon dioxide necessary for the termination of global glaciation, *Nature*, **429**, 646–649.

Pollack, H.N., 1986, Cratonization and thermal evolution of the mantle, *Earth and Planetary Science Letters*, **80**, 175–182.

Pollack, J.B., O. Hubickyj, P. Bodenheimer, J.J. Lissauer, M. Podolak, and Y. Greenzweig, 1996, Formation of the giant planets by concurrent accretion of solids and gas, *Icarus*, **124**, 62–85.

Pollard, D., and J.F. Kasting, 2004, Climate-ice sheet simulations of Neoproterozoic glaciation before and after collapse to snowball Earth, in *Multidisciplinary Studies Exploring Extreme Proterozoic Environment Conditions*, G. Jenkins, C. McKay, M. McMenamin, and L. Sohl, eds., American Geophysical Union, Washington, D.C., pp. 91–105.

Pollard, D., and J.F. Kasting, 2005, Snowball Earth: A thin-ice model with flowing sea glaciers, *Journal of Geophysical Research*, **110**, C07010.

Pörtner, H.O., M. Langenbuch, and B. Michaelidis, 2005, Synergistic effects of temperature extremes, hypoxia, and increases in CO_2 on marine animals: From Earth history to global change, *Journal of Geophysical Research—Oceans*, **110**, C09S10.

Press, F., and R. Siever, 2001, *Understanding Earth*, 3rd ed., W.H. Freeman, New York, 573 pp.

Rampino, M.R., and S. Self, 1992, Volcanic winter and accelerated glaciation following the Toba super-eruption, *Nature*, **359**, 50–52.

Ravizza, G., R.N. Norris, J. Blusztajn, and M.-P. Aubry, 2001, An osmium isotope excursion associated with the late Paleocene thermal maximum: Evidence of intensified chemical weathering, *Paleoceanography*, **16**, 155–163.

Raymo, M.E., and W.F. Ruddiman, 1993, Tectonic forcing of late Cenozoic climate, *Nature*, **361**, 117–122.

Ren, Y., E. Stutzman, R.D. van der Hilst, and J. Besse, 2007, Understanding seismic heterogeneities in the lower mantle beneath the Americas from seismic tomography and plate tectonic history, *Journal of Geophysical Research—Solid Earth*, **112**, B01302.

Ribe, N.M., and U.R. Christensen, 1999, The dynamical origin of Hawaiian volcanism, *Earth and Planetary Science Letters*, **171**, 517–531.

Ricardo, A., M.A. Carrigan, A.N. Olcutt, and S.A. Benner, 2004, Borate minerals stabilize ribose, *Science*, **303**, 196.

Richards, M.A., W.-S. Yang, J.R. Baumgardner, and H.-P. Bunge, 2001, Role of a low-viscosity zone in stabilizing plate tectonics: Implications for comparative terrestrial planetology, *Geochemistry Geophysics Geosystems*, **2**, 2000GC000115. Correction published July 9, 2002.

Richter, F.M., 1973, Dynamical models for sea-floor spreading, *Reviews of Geophysics and Space Physics*, **11**, 223–287.

Richter, F.M., 1985, Models for the Archean thermal regime, *Earth and Planetary Science Letters*, **73**, 350–360.

Richter, F.M., and D.P. McKenzie, 1984, Dynamical models for melt segregation from a deformable matrix, *Journal of Geology*, **93**, 729–740.

Righter, K., M.J. Drake, and G. Yaxley, 1997, Prediction of siderophile element metal-silicate partition coefficients to 20 GPa and 2800 degrees C: The effects of pressure, temperature, oxygen fugacity, and silicate and metallic melt compositions, *Physics of the Earth and Planetary Interiors*, **100**, 115–134.

Röhl, U., T.J. Bralower, R.D. Norris, and G. Wefer, 2000, New chronology for the late Paleocene thermal maximum and its environmental implications, *Geology*, **28**, 927–930.

Romanowicz, B., 2008, Using seismic waves to image Earth's internal structure, *Nature*, **451**, 266-268.

Rose, W.I., and C.A. Chesner, 1987, Dispersal of ash in the great Toga eruption, 75 ka, *Geology*, **15**, 913–917.

Rothman, D.H., J.M. Hayes, and R.E. Summons, 2003, Dynamics of the Neoproterozoic carbon cycle, *Proceedings of the National Academy of Sciences*, **100**, 8124–8129.

Rowley, D.B., 2002, Rate of plate creation and destruction: 180 Ma to present, *Geological Society of America Bulletin*, **114**, 927–933.

Royer, J.-Y., and R.G. Gordon, 1997, The motion and boundary between the Capricorn and Australian plates, *Science*, **277**, 1268–1274.

Royer, D.L., R.A. Berner, and D.J. Beerling, 2001, Phanerozoic atmospheric CO_2 change: Evaluating geochemical and paleobiological approaches, *Earth-Science Review*, **54**, 349–392.

Rumble, D., J.G. Liou, and B.-M. Jahn, 2003, Continental crust subduction and UHP metamorphism, in *The Crust, Treatise on Geochemistry*, **3**, R.L. Rudnick. ed., Elsevier, Amsterdam, pp. 293–319.

Sakellariou, A., T.J. Sawkins, T.J. Senden, and A. Limaye, 2004, X-ray tomography for mesoscale physics applications, *Physica A*, **339**, 152–158.

Salters, V., and A. Stracke, 2004, Composition of the depleted mantle, *Geochemistry Geophysics Geosystems*, **5**, Q05004.

Satake, K., 2007, Tsunamis, in *Treatise on Geophysics*, **4**, H. Kanamori, ed., Elsevier, New York, pp. 483–511.

Scarth, A., 2002, *La Catastrophe: The Eruption of Mount Pelee, the Worst Volcanic Disaster of the 20th Century*, Oxford University Press, Oxford, 256 pp.

Scherstén, A., T. Elliott, C. Hawkesworth, and M. Norman, 2004, Tungsten isotope evidence that mantle plumes contain no contribution from the Earth's core, *Nature*, **427**, 234–237.

Scholz, C.H., 1990, *The Mechanics of Earthquakes and Faulting*, Cambridge University Press, New York, 439 pp.

Schoonen, M., A. Smirnov, and C. Cohn, 2004, A perspective on the role of minerals in prebiotic synthesis, *Ambio*, **33**, 539–551.

Schopf, J.W., 2006, Fossil evidence of Archaean life, *Philosophical Transactions of the Royal Society*, London, **B361**, 869–885.

Schopf, J.W., A.B. Kudryavtsev, D.G. Agresti, T.J. Wdowiak, and A.D. Czaja, 2002, Laser-Raman imagery of Earth's earliest fossils, *Nature*, **416**, 73–76.

Sepkoski, J.J., 1984, A kinetic model of Phanerozoic taxonomic diversity. 3. Post-Paleozoic families and mass extinctions, *Paleobiology*, **10**, 246–267.

Shen, G.Y., V.B. Prakapenka, M.L. Rivers, and S.R. Sutton, 2004, Structure of liquid iron at pressures up to 58 GPa, *Physical Review Letters*, **92**, 185701.

Sigurdsson, H., and S. Carey, 1989, Plinian and co-ignibrite tephra fall from the 1815 eruption of Tambora volcano, *Bulletin of Volcanology*, **51**, 243–270.

Sillitoe, R.H., and J.W. Hedenquist, 2003, Linkages between volcanotectonic settings, ore-fluid compositions, and epithermal precious metal deposits, in *Volcanic, Geothermal, and Ore-Forming Fluids: Rulers and Witnesses of Processes Within the Earth*, S.F. Simmons, and I. Graham, eds., Society of Economic Geologists, Special Publication 10, pp. 315–343.

Simkin, T., and R.S. Fiske, 1983, *Krakatau 1883—The Volcanic Eruption and Its Effects*, Smithsonian Institution Press, Washington, D.C., 470 pp.

Sims, K.W., S.J. Goldstein, J. Blichert-Toft, M.R. Perfit, P. Kelemen, D.J. Fornari, P. Michael, M.T. Murrell, S.R. Hart, D.J. DePaolo, G. Layne, L. Ball, M. Jull, and J. Bender, 2002, Chemical and isotopic constraints on the generation and transport of magma beneath the East Pacific Rise, *Geochimica et Cosmochimica Acta*, **66**, 3481–3504.

Sleep, N.H., 2002, Self-organization of crustal faulting and tectonics, *International Geology Review*, **44**, 83–96.

Sleep, N.H., 2005, Evolution of the continental lithosphere, *Annual Review of Earth and Planetary Sciences*, **33**, 369–393.

Sleep, N.H., and D.K. Bird, 2007, Niches of the pre-photosynthetic biosphere and geologic preservation of Earth's earliest ecology, *Geobiology*, **5**, 101–117.

Sleep, N.H., K.J. Zahnle, J.F. Kasting, and H.J. Morowitz, 1989, Annihilation of ecosystems by large asteroid impacts on the early Earth, *Nature*, **342**, 139–142.

Sonnenthal, E., N. Spycher, and C. Haukwa, 2001, Thermal-hydrologic-chemical effects on seepage and potential seepage chemistry, in *Supplemental Science and Performance Analyses, Volume 1, Scientific Bases and Analyses, REV 00*, available at <http://www.osti.gov/bridge/servlets/purl/784533-H4LgEw/webviewable/784533.PDF>.

Spiegelman, M., and P.B. Kelemen, 2003, Extreme chemical variability as a consequence of channelized melt transport, *Geochemistry Geophysics Geosystems*, **4**, 1055.

Spiegelman, M., P.B. Kelemen, and E. Aharonov, 2001, Causes and consequences of flow organization during melt transport: The reaction infiltration instability in compactible media, *Journal of Geophysical Research—Solid Earth*, **106**, 2061–2077.

Squyres, S.W., J.P. Grotzinger, R.E. Arvidson, J.F. Bell, W. Calvin, P.R. Christensen, B.C. Clark, J.A. Crisp, W.H. Farrand, K.E. Herkenhoff, J.R. Johnson, G. Klingelhofer, A.H. Knoll, S.M. McLennan, H.Y. McSween, R.V. Morris, J.W. Rice, R. Rieder, and L.A. Soderblom, 2004, In situ evidence for an ancient aqueous environment at Meridiani Planum, Mars, *Science*, **306**, 1709–1714.

Stanley, S.M., 2007, *An Analysis of the History of Marine Animal Diversity*, Paleobiology Memoir 4, 55 pp.

Steefel, C.I., D.J. DePaolo, and P.C. Lichtner, 2005, Reactive transport modeling: An essential tool and a new research approach for the Earth sciences, *Earth and Planetary Science Letters*, **240**, 539–558.

Stevens, S.H., V.K. Kuuskraa, and J. Gale, 2000, Sequestration of CO_2 in depleted oil and gas fields: Global capacity and barriers to overcome, in *Proceedings of the 5th International Conference on Greenhouse Gas Control Technologies (GHGT5)*, Cairns, Australia, August, 13-16, 2000.

Stevenson, D.J., 1987, Origin of the moon—The collision hypothesis, *Annual Review of Earth and Planetary Sciences*, **15**, 271–315.

Stixrude, L., and J.M. Brown, 1998, The Earth's core, *Reviews in Mineralogy*, **37**, 261–283.

Stixrude, L., and B. Karki, 2005, Structure and freezing of $MgSiO_3$ liquid in Earth's lower mantle, *Science*, **310**, 297–299.

Stoll, H.M., 2006, The Arctic tells its story, *Nature*, **441**, 579–581.

Stothers, R.B., 1984, The great Tambora eruption of 1815 and its aftermath, *Science*, **224**, 1191–1198.

Su, Y.-J., 2002, *Mid-ocean Ridge Basalt Trace Element Systematics: Constraints from Database Management, ICPMS Analysis, Global Data Compilation, and Petrologic Modeling*, Ph.D. Dissertation, Columbia University, 569 pp.

Su, W.J., R.L. Woodward, and A.M. Dziewonski, 1994, Degree-12 shear velocity model of mantle heterogeneity, *Journal of Geophysical Research*, **99**, 6945–6980.

Syvitski, J.P.M., S.D. Peckham, R.D. Hilberman, and T. Mulder, 2003, Predicting the terrestrial flux of sediment to the global ocean: A planetary perspective, *Sedimentary Geology*, **162**, 5–24.

Syvitski, J.P.M., C.J. Vörösmarty, A.J. Kettner, and P. Green, 2005, Impact of humans on the flux of terrestrial sediment to the global coastal ocean, *Science*, **308**, 376–380.

Tarduno, J.A., D.B. Brinkman, P.R. Renne, R.D. Cottrell, H. Scher, and P. Castillo, 1998, Evidence for extreme climatic warmth from late Cretaceous arctic vertebrates, *Science*, **282**, 2241–2243.

Tarduno, J.A., R.D. Cottrell, M.K. Watkeys, and D. Bauch, 2007, Geomagnetic field strength 3.2 billion years ago recorded by single silicate crystals, *Nature*, **446**, 657–660.

Tatsumi, Y., and S. Eggins, 1995, *Subduction Zone Magmatism*, Blackwell Science, Cambridge, MA, 211 pp.

Thordarson, T., and S. Self, 2003, Atmospheric and environmental effects of the 1783–1784 Laki eruption: A review and reassessment, *Journal of Geophysical Research*, **108**, doi:10.1029/2001JD002042.

Tian, F., O.B. Toon, A.A. Pavlov, and H. DeSterck, 2005, A hydrogen-rich early Earth atmosphere, *Science*, **308**, 1014–1017.

Toksöz, M.N., A.F. Shakal, and A.J. Michael, 1979, Space-time migration of earthquakes along the North Anatolian fault zone and seismic gaps, *Pure and Applied Geophysics*, **117**, 1258–1270.

Touma, J., and J. Wisdom, 1998, Resonances in the early evolution of the Earth-Moon system, *The Astronomical Journal*, **115**, 1653–1663.

Trampert, J., and R.D. van der Hilst, 2005, Towards a quantitative interpretation of global seismic tomography, in *Earth's Deep Mantle: Structure, Composition, and Evolution*, R.D. van der Hilst, J. Bass, J. Matas, and J. Trampert, eds., Geophysical Monograph Series 160, pp. 47–62.

Triebs, A., 1936, Chlorophyll and hemin derivatives in organic mineral substances, *Angewandte Chemie*, **49**, 682–686.

Tsuchiya, T., R.M. Wentzcovitch, C.R.S. da Silva, and S. de Gironcoli, 2006, Spin transition in magnesiowüstite in Earth's lower mantle, *Physical Review Letters*, **96**, 198501.

Turcotte, D.L., and E.R. Oxburgh, 1967, Finite amplitude convective cells and continental drift, *Journal of Fluid Mechanics*, **28**, 29–42.

Turcotte, D.L., R. Shcherbakov, and J.B. Rundle, 2007, Complexity and earthquakes, in *Earthquake Seismology, Treatise on Geophysics*, **4**, H. Kanamori, ed., Elsevier, New York, pp. 675–700.

Valley, J.W., W.H. Peck, E.M. King, and S.A. Wilde, 2002, A cool early Earth, *Geology*, **30**, 351–354.

van der Hilst, R.D., M.V. de Hoop, P. Wang, S.-H. Shim, P. Ma, and L. Tenorio, 2007, Seismostratigraphy and thermal structure of Earth's core-mantle boundary region, *Science*, **315**, 1813–1817.

Van Keken, P.E., E.H. Hauri, and C.J. Ballentine, 2002, Mantle mixing: The generation, preservation and destruction of chemical heterogeneity, *Annual Review of Earth and Planetary Sciences*, **30**, 493–525.

Veizer, J., Y. Godderis, and L.M. François, 2000, Evidence for decoupling of atmospheric CO_2 and global climate during the Phanerozoic eon, *Nature*, **408**, 698–701.

Voight, B., 1990, The 1985 Nevado del Ruiz volcano catastrophe: Anatomy and retrospection, *Journal of Volcanological and Geothermal Research*, **44**, 349–386.

von Huene, R., and D.W. Scholl, 1991, Observations at convergent margins concerning sediment subduction, subduction erosion, and the growth of continental crust, *Reviews of Geophysics*, **29**, 279–316.

Vörösmarty, C.J., M. Meybeck, B. Fekete, K. Sharma, P. Green, and J.P.M. Syvitski, 2003, Anthropogenic sediment retention: Major global impact from registered river impoundments, *Global and Planetary Change*, **39**, 169–190.

Wächtershäuser, G., 1988, Before enzymes and templates: Theory of surface metabolism, *Microbiology Review*, **52**, 452–484.

Walker, J.C.G., P.B. Hays, and J.F. Kasting, 1981, A negative feedback mechanism for the long-term stabilization of Earth's surface temperature, *Journal of Geophysical Research*, **86**, 9776–9782.

Watson, E.B., J.B. Thomas, and D.J. Cherniak, 2007, [40]Ar retention in the terrestrial planets, *Nature*, **449**, 299–304.

Weidenschilling, S.J., 1997, The origin of comets in the solar nebula: A unified model, *Icarus*, **127**, 290–306.

Weiss, B.P., J.L. Kirschvink, F.J. Baudenbacher, H. Vali, N.T. Peters, F.A. Macdonald, and J.P. Wikswo, 2000, A low temperature transfer of ALH84001 from Mars to Earth, *Science*, **290**, 791–795.

West, M., J.J. Sanchez, and S.R. McNutt, 2005, Periodically triggered seismicity at Mount Wrangell, Alaska, after the Sumatra earthquake, *Science*, **308**, 1144–1146.

Wiegel, J., and M.W.W. Adams, eds., 1998, *Thermophiles: The Keys to Molecular Evolution and the Origin of Life*, Taylor and Francis, London, 346 pp.

Wignall, P.B., and R.J. Twitchett, 1996, Oceanic anoxia and the end Permian mass extinction, *Science*, **272**, 1155–1158.

Wilde, S.A., J.W. Valley, W.H. Peck, and C.M. Graham, 2001, Evidence from detrital zircons for the existence of continental crust and oceans on the Earth 4.4 Gyr ago, *Nature*, **409**, 175–178.

Wilhelms, D.E., 1987, *The Geologic History of the Moon*, USGS Professional Paper 1348, U.S. Government Printing Office, 302 pp.

Willett, S.D., 1999, Orography and orogeny: The effects of erosion on the structure of mountain belts, *Journal of Geophysical Research*, **104**, 28,957–28,981.

Williams, Q., and E.J. Garnero, 1996, Seismic evidence for partial melt at the base of Earth's mantle, *Science*, **273**, 1528.

Williams, Q., and R.J. Hemley, 2001, Hydrogen in the deep Earth, *Annual Review of Earth and Planetary Sciences*, **29**, 365–418.

Wilson, P.A., and R.D. Norris, 2001, Warm tropical ocean surface and global anoxia during the mid-Cretaceous period, *Nature*, **412**, 425–429.

Woese, C.R., 1977, Endosymbionts and mitochondrial origins, *Journal of Molecular Evolution*, **10**, 93–96.

Wood, B.J., M.J. Walter, and J. Wade, 2006, Accretion of the Earth and segregation of its core, *Nature*, **441**, 825–833.

Zachos, J., M. Pagani, L. Sloan, E. Thomas, and K. Billups, 2001, Trends, rhythms, and aberrations in global climate 65 Ma to present, *Science*, **292**, 686–693.

Zachos, J.C., U. Rohl, S.A. Schellenberg, A. Sluijs, D.A. Hodell, D.C. Kelly, E. Thomas, M. Nicolo, I. Raffi, L.J. Lourne, H. McCarren, and D. Kroon, 2005, Rapid acidification of the ocean during the Paleocene-Eocene thermal maximum, *Science*, **308**, 1611–1615.

Zahnle, K.J., 2006, Earth's earliest atmosphere, *Elements*, **2**, 217–222.

Zahnle, K.J., and N.H. Sleep, 1997, Impacts and the early evolution of life, in *Comets and the Origin and Evolution of Life*, P. Thomas, C.F. Chyba, and C.P. McKay, eds., Springer-Verlag, New York, pp. 175–208.

Zahnle, K., N. Arndt, C. Cockell, A. Halliday, E. Nisbet, F. Selcis, and N. Sleep, 2007, Emergence of a habitable planet, *Space Science Reviews*, **24**, 35–78.

Zandt, G., H. Gilbert, T.J. Owens, M. Ducea, J. Saleeby, and C.H. Jones, 2004, Active foundering of a continental arc root beneath the southern Sierra Nevada in California, *Nature*, **431**, 41–46.

Zatman, S., R.G. Gordon, and M.A. Richards, 2001, Analytic models for the dynamics of diffuse oceanic plate boundaries, *Geophysical Journal International*, **145**, 145–156.

Zinner, E.K., 2003, Presolar grains, in *Meteorites, Comets, and Planets, Treatise on Geochemistry*, **1**, A.M. Davis, ed., Elsevier-Pergamon, Oxford, pp. 17–40.

Appendix A

Biographical Sketches of Committee Members

Donald J. DePaolo, *Chair*, is the Class of 1951 Professor of Geochemistry at the University of California, Berkeley; director of the Center for Isotope Geochemistry; and director of the Earth Sciences Division at the Lawrence Berkeley National Laboratory. He received his Ph.D. in geology from the California Institute of Technology. Dr. DePaolo's research interests are in the application of radiogenic isotope geochemistry and principles of physics and chemistry to problems in geology, geophysics, and environmental science. He has served on several advisory committees concerned with the geosciences, including the National Research Council's (NRC's) Board on Earth Sciences and Resources and its Geodynamics Committee, and has chaired numerous professional society, advisory, and university visiting committees. Dr. DePaolo is the recipient of many awards for his contributions to the geochemical and geophysical sciences, including the American Geophysical Union's Macelwane Award, the Geochemical Society's F.W. Clark Medal, the Geological Society of America's Arthur L. Day Medal, and the European Association of Geochemistry's Harold Urey Medal. He is a member of the National Academy of Sciences and the American Academy of Arts and Sciences.

Thure E. Cerling is Distinguished Professor of Geology and Geophysics and Distinguished Professor of Biology at the University of Utah. He received his Ph.D. in geology from the University of California, Berkeley. His research concerns near-surface processes and the geological record of ecological change. Of particular interest are isotope physiology and paleodiets of mammals, soils as indicators of climatological and ecological change over geological timescales, and landscape evolution over the last several million years. Dr. Cerling has served on several NRC solid-earth committees, including the Board on Earth Sciences and Resources, the Geodynamics Committee, and the U.S. National Committee for the International Union for Quaternary Research. He is member of the U.S. Nuclear Waste Technical Review Board, a fellow of both the American Association for the Advancement of Science and of the Geological Society of America, and a member of the National Academy of Sciences.

Sidney R. Hemming is an associate professor in the Department of Earth and Environmental Sciences at Columbia University. She earned her Ph.D. in geology from the State University of New York, Stony Brook. Her research focuses on paleoceanography, paleoclimate, tracer studies, and geochemistry of sedimentary rocks. Several of her current projects deal with reconstructing ocean circulation patterns at different times, including periods of abrupt climate change. Other projects aim to understand the role of ice sheets in regional and global climate change, including studies of the North Atlantic and circum-Antarctic oceans and Mono Lake. Dr. Hemming is deputy director (education liaison) of the Cooperative Institute for Climate Applications and Research, a research partnership on climate variability and change sponsored by the National Oceanic and Atmospheric Administration and Columbia. She is a member of the American Geophysical Union and the

Geochemical Society and is on the editorial board of the journal *Chemical Geology*.

Andrew H. Knoll is Fisher Professor of Natural History and curator of the Paleobotanical Collections, Botanical Museum, at Harvard University. His geology Ph.D. was also from Harvard University. His research interests are in Precambrian biological and geological evolution, early animal diversification, vascular plant evolution, and the relationship between evolution and environmental change in Earth history. He also has an interest in astrobiology and was a member of the rover science team in the National Aeronautics and Space Administration's 2003 mission to Mars. Dr. Knoll has received several awards for his scientific achievements, including the Paleontological Society's Medal and Charles Schuchert Award and the Society for Sedimentary Geology's Raymond C. Moore Medal. He has served on Earth and space science advisory groups, including the NRC Space Studies Board and the Board on Earth Sciences and Resources. He is a member of the National Academy of Sciences.

Frank M. Richter is Sewell Avery Distinguished Service Professor in the Department of Geophysical Sciences at the University of Chicago. He received his Ph.D. from the University of Chicago. Dr. Richter's research spans both geophysics and geochemistry and includes investigations of mantle convection, thermal evolution of Earth, isotopic dating, pore-water chemistry in sediments, and melt segregation and chemical diffusion in molten rock systems. Both lines of research have led to professional society awards, including the American Geophysical Union's Bowen Award and the Geological Society of America's Wollard Award. Dr. Richter has served on numerous NRC solid-earth science committees, including the Board on Earth Sciences and Resources, Geodynamics Committee (chair), Committee on Seismology, and Committee on Basic Research Opportunities in the Earth Sciences. He is a member of the National Academy of Sciences.

Leigh H. Royden is professor of geology and geophysics and chair of the Program in Geology and Geochemistry at the Massachusetts Institute of Technology. She received her Ph.D. from the same institution. Dr. Royden's research interests include regional geology and geophysics and the mechanics of large-scale continental deformation. She has received the Geological Society of America's Donath Medal and a Presidential Young Investigator Award. She has served on the Council of the Geological Society of America and is a former member of the NRC Geodynamics Committee and Committee on Basic Research Opportunities in the Earth Sciences. She is a fellow of both the Geological Society of America and the American Geophysical Union.

Roberta L. Rudnick is a professor in the Department of Geology at the University of Maryland. Prior to joining the faculty in 2000, she spent six years as a professor at Harvard University and several years as a research scientist at the Max Planck Institute für Chemie in Mainz, Germany. Dr. Rudnick received her Ph.D. from the Research School of Earth Sciences at Australian National University. Her research focuses on the origin and evolution of the continents, particularly the lower continental crust and the underlying mantle lithosphere. In addition to her research, she is a councillor for the Mineralogical Society of America and editor-in-chief of *Chemical Geology*. She is a fellow of the American Geophysical Union, the Geological Society of America, and the Mineralogical Society of America and has been a distinguished lecturer for the latter society.

Lars Stixrude is a professor of geophysics and mineral physics at the University of Michigan. He received his Ph.D. in geophysics at the University of California, Berkeley. Dr. Stixrude investigates the physics of Earth at an atomic level. Predictions of material physics at conditions of Earth's interior, based on theoretical and laboratory investigations, provide insight into magma generation and transport, the seismic structure of the mantle and core, and the state of water in the deep interior. He is a member of the steering committee for the Cooperative Institute for Deep Earth Research, which is developing an intellectual framework to improve communication among scientists in different disciplines studying the dynamics of Earth's interior. He is a recipient of the American Geophysical Union's James B. Macelwane Medal and a fellow of both the Mineralogical Society of America and the American Geophysical Union.

James S. Trefil is Clarence J. Robinson Professor of Physics at George Mason University. He earned his Ph.D. in theoretical physics from Stanford University. In addition to his research on particle physics, field theory, astrophysics, and cosmology, he has strong interests in teaching science to nonscientists. His course and textbook series on achieving scientific literacy is used in approximately 200 colleges and universities, and he has written numerous articles and books for general audiences. His "Ask Mr. Science" column ran in *USA Today* from 1996 to 1999. He has also served as a science commentator and member of the Science Advisory Board for National Public Radio. Dr. Trefil is a fellow of the American Physical Society, the American Association for the Advancement of Science, and the World Economic Forum. He is a recipient of the American Institute of Physics Andrew Gemant Award for sustained contributions in bridging the gap between science and society.

Appendix B

Acronyms and Abbreviations

AU	astronomical unit
CAI	calcium-, aluminum-rich inclusion
Ga	billion years ago
GPS	Global Positioning System
InSAR	interferometric synthetic aperture radar
M	magnitude
Ma	million years ago
THMC	thermal-hydrological-mechanical-chemical
WFPC2	Wide Field and Planetary Camera 2